Homework and Test Questions for Introductory Physics Teaching

Arnold B. Arons
UNIVERSITY OF WASHINGTON

[*Warning from the author: It is unwise to subject students to questions or problems out of this book without reading the preface and following the advice given concerning rewriting of the text to match the background and vocabulary familiar to the students.*]

JOHN WILEY & SONS, INC.

New York · Chichester · Brisbane · Toronto · Singapore

ACQUISITIONS EDITOR   Cliff Mills

MARKETING MANAGER   Catherine Faduska

PRODUCTION EDITORS   Deborah Herbert and Lenore Belton

DESIGNER   Ann Renzi

MANUFACTURING MANAGER   Andrea Price

ILLUSTRATION   Anna Melhorn

This book was set in Times Roman by the author and
printed and bound by Hamilton Printing Company.

Library of Congress Cataloging in Publication Data:
Arons, A. B. (Arnold B.)
  Homework and test questions for introductory physics teaching /
  Arnold B. Arons.
    p.   cm.
  Includes bibliographical references.
  ISBN 0-471-30931-1
  1. Physics--Study and teaching. 2. Physics--Study and teaching-
  -Examinations, questions, etc. 3. Physics--Study and teaching-
  - Problems, exercises, etc. 4. Homework. I. Title.
  QC30.A76 1990 Suppl.
  530'.076--dc20   93-48576
                                                   CIP

Printed in the United States of America

10 9 8 7 6 5 4 3 2 1

Preface

People have now-a-days got a strange opinion that everything should be taught by lectures. Now, I cannot see that lectures can do as much good as reading the book from which the lectures are taken.

— Samuel Johnson

'You damn sadist,' said mr. cummings, 'You try to make people think.'

— Ezra Pound, Canto 89

This collection of questions and problems is supplementary to the questions and problems found in *A Guide to Introductory Physics Teaching* (Wiley, New York, 1990). In the *Guide*, questions and problems, illustrating approaches that lead the student into physical experiences and into sequences of thinking and reasoning that help penetrate learning difficulties, are placed in appendices to some chapters and are also embedded in the text itself where research protocols are described. Readers should realize that questions that illuminated student preconceptions, misconceptions, and other learning difficulties in the research investigations are themselves very useful when embedded in homework or in tests. To conserve space and limit cost, only a few of the questions and problems suggested in the *Guide* are repeated here, and readers are therefore referred to the *Guide* for these prior materials.

The great majority of the questions and problems in this collection have been used and tested either by the author in his own courses or by colleagues who have provided ideas based on their own experience. Some of the questions stem from research experience in our Physics Education Research Group. Many of the questions are especially well suited for use in group discussion and cooperative learning formats.

This collection is confined to a very basic level of subject matter common to the great majority of introductory physics courses; more advanced subjects and concepts have been deliberately excluded. Within this range of subject

iii

matter, however, the questions range from extremely simple and fundamental to fairly sophisticated. In the latter, fairly extensive guidance is provided. No attempt is made to duplicate conventional types of numerical end-of-chapter problems readily available in existing texts. The texts are replete with excellent examples. These are a necessary and intrinsic part of our teaching, and this collection is not meant to deprecate or diminish their role. The intent is to fill what researches on teaching and learning show to be gaps in the existing structure. In those instances where fairly conventional questions are being presented, they are generally amplified by addition of Socratic questioning that helps penetration of the question by students who do not otherwise get a start.

These questions and problems spread over a variety of modes that emerge as essential components in the learning and understanding of physics. Among these modes are forming and applying basic concepts; operational definition; verbalization; connection of abstractions to everyday experience; translating between various representations (e.g., verbal to symbolic or to graphic and the reverse); asking questions; formulating problems; visualizing outcomes in the abstract (hypothetico-deductive reasoning); discriminating between observation and inference; checking for internal consistency; and interpreting results. Students need extensive practice in these modes of thinking and reasoning, and under supply of such opportunities is one of the principal shortcomings of conventional texts and much physics instruction. This collection is a limited attempt to bridge some of the existing gaps, but I do not pretend that it is either complete or final. Much will be learned through research in the coming years, and much can be added by practicing teachers who bring their own varied imaginations to the building of more comprehensive and more effective collections. No one looks forward more eagerly than I to the additions that invariably emerge when fresh imaginations supplement the limited output of a single individual or a small initial group.

Users of this collection should understand, however, that, regardless of the prior use indicated, it is unwise to take the questions and problems exactly as they are articulated in this book and in the *Guide* and pass them, unaltered and unrefined, to their students. The items that are presented here are meant to be illustrative — to serve as nuclei, to provide hints and ideas for teachers to view as starting points for transformation to their own framework (or for entirely new questions of their own) rather than as fixed end points for immediate use. In other words, the suggested items should be reworked and reworded to fit smoothly and consistently with the presentations a teacher has used and, even more importantly, to fit the *vocabulary a teacher has been employing*. If these conditions are not met, questions, however well conceived in principle, will be meaningless and unintelligible to the students. The necessary transmutation must be effected by each individual teacher in the context in which he or she operates.

Because of this need for reworking and transmutation, and also to keep down the size and cost of this volume, I have not included my own answers and solutions. These are, in fact, irrelevant. Teachers should not make use of questions about which they have doubts of any kind; they should rework any nucleus of an idea that appeals to them into a form that fits what they have been doing with their students. My own phraseology, for example, which may be clear to my own students because of the locutions I have used from the very beginning, may well be ambiguous in some other context and may require substantial revision for use by another teacher. Such revision, of necessity, remains up to the individual teacher.

It is my firm belief that testing along many of the lines illustrated in this collection (in addition to utilizing the more conventional end-of-chapter type problems) is essential for impelling students toward firmer concept formation and genuine understanding of the physics. In the absence of such testing, many students tend to memorize, without understanding, approaches to type-problem solving that yield sufficient partial credit for an adequate grade. They learn to avoid the labor of thought that yields the levels of understanding to which we (the teachers) persistently render lip service, and they emerge from our courses with the misapprehensions and lack of understanding so painfully revealed in the accumulating researches. Some of our students have never solved a problem completely correctly in their entire experience in physics, and we are vulnerable to the charge of encouraging the formation of what I have come to call "partial credit minds."

In addition to questions and problems, I have included two other items that might be of use to some teachers. One is a set of statements of learning objectives that were formulated in our group at a time when we were operating summer institutes for high school physics teachers. (The participants were working in pairs in self-paced mode, and the statements of learning objectives were useful, even necessary, guides to help them achieve the depth of understanding we considered essential for teaching.) The other item consists of several examples of term paper assignments for an introductory physics course. We had found that term paper assignments were largely ineffective if beginning students departed into large subjects without some reasonable degree of constraint, guidance, and concentration. The examples given are intended to show how effective assignments, which provide guidance and choice without excessive constraint, can be constructed. They are intended as helpful models and suggestions rather than as items to be taken and implemented as they stand.

I am indebted to many sources for ideas entering into this collection. Some originated so long ago that I can no longer separate my own contribution from those of colleagues who worked with me over the years. I can only express gratitude to all who helped criticize and refine the questions that evolved in our courses. In more recent time, I am especially indebted to my colleague Philip Peters for a variety of fertile ideas and for review of many items in this collection.

Arnold B. Arons
Seattle, Washington
November 1993

Contents

CHAPTER 1

Scaling and Ratio Reasoning

Note to the instructor: Most of the following questions are designed to lead students into thinking about scaling, and about functional relationships, in terms of ratios rather than through substitution in formulas. Probably because of lack of practice and unfamiliarity with this mode of thought, many students exhibit strong resistance to ratio reasoning and strive to convert it into formula substitution, avoiding the reasoning. Since such ratio reasoning, however, underlies virtually all estimating, scaling, and thinking in terms of orders of magnitude, its cultivation is highly desirable and important in student cognitive development. In this chapter, the range of context has been highly restricted, and the questions are essentially illustrative. Teachers would be well advised to provide many opportunities for such reasoning in every possible context of subject matter, using the basic pattern being illustrated if it fits the teacher's own mode of operation. To register their importance, however, it is essential to *test* on scaling and ratio reasoning. It is the usual *absence* of such testing that has left many students where they are and has helped crystallize their frequently massive, even fierce, resistance.

1.1 In a uniformly accelerated rectilinear motion, starting from rest, a particle undergoes a displacement of 3.36 m in a time interval $(\Delta t)_1$. What would have been the time interval $(\Delta t)_2$ for a displacement over the first 1.28 m in this same history? [Do not substitute in formulas. Select the relevant functional relationship between displacement and time interval and then express $(\Delta t)_2$ directly in terms of $(\Delta t)_1$ and an appropriate numerical ratio based on the functional relation.]

1

1.2 From the kinematic relations for uniformly accelerated rectilinear motion with negligible frictional resistance, find the expression for how high an object rises when it is thrown vertically upward with an initial velocity v_0. Explain your steps of reasoning in making the derivation.

(a) The acceleration due to gravity at the surface of the moon is about 1/6 that at the surface of the earth. For a given initial vertical velocity v_0, how much higher will the object rise if thrown upward on the moon than if thrown upward on the earth? Explain your reasoning. (Use *ratio reasoning* based on the functional relation you have obtained; do *not* substitute in the formula.)

(b) Suppose we increase the initial velocity by a factor of 2.6. By what factor will the height of rise increase at each location? Explain your reasoning. (Use ratio reasoning based on the functional relation; do not substitute in the formula.)

1.3 It is an established fact that the breaking strength of a stretched wire is directly proportional to the cross-sectional area of the wire. ["Breaking strength" is defined as the total stretching force (tension) at which the wire breaks.] Suppose a brass wire 0.56 mm in diameter is just approaching the breaking point under a tension of 85 N.

(a) Suppose we wish to suspend a block with a mass of 24 kg on a brass wire as shown in the diagram. Explaining your reasoning, calculate the diameter of the wire that would be necessary to hold the block without breaking. (Pay careful attention to the units you use and to the distinction between weight and mass.)

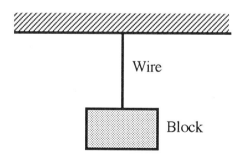

(b) Is the diameter you calculated in part (a) a minimum value, a maximum value, some intermediate value? Explain your reasoning.

(c) Suppose the *length* of the suspending wire is now increased by a factor of 1.6 without any change in the cross-sectional area. Will the maximum weight the wire will support increase, decrease, or remain unchanged? Explain your reasoning.

1.4 A concrete block of the shape shown hangs from a thin metal strip having the cross-sectional shape indicated in the inset on the left of the diagram. Suppose the block is increased in size through scaling up each of its three linear dimensions by a factor of 2.10 without change in the material of which it is composed.

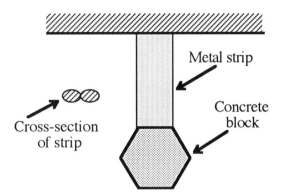

(a) By what *factor* will the weight of the block be increased because of the change in dimensions? (Calculate the numerical value and explain your reasoning briefly.)

(b) By what scale factor must the linear dimensions of the cross section of the suspending strip be changed to compensate for the change in weight of the block, making sure that the strip does not break? Explain your reasoning briefly. (Remember that the breaking strength of the strip is proportional to its cross-sectional *area*, not its volume.)

1.5 Suppose that in a certain atmospheric situation, a falling raindrop, having an initial mass of 0.010 g, collects smaller droplets by collision at such a rate that, after 10 s, it has a mass of 0.020 g and a diameter of 3.4 mm.

(a) If the droplet continues to gain mass at the same average rate as it continues to fall, what will be its diameter at the end of the next 10 s? (Use scaling and ratio reasoning, not substitution in formulas, and explain what you are doing.)

(b) Is 3.4 mm a reasonable or unreasonable *order of magnitude* for the diameter of a droplet with a mass of 0.020 g? (Make a crude, quick estimate — not an elaborate calculation — and explain your reasoning.)

1.6 A statue is to be scaled down, without distorting its shape, by changing its total volume from 1.25 m³ to 0.37 m³. Explain your reasoning in each of the following calculations.

(a) If the height of the original statue is 250 cm, calculate the height of the smaller model.

(b) If the circular base of the original statute has a circumference of 45 cm, calculate the circumference of the scaled-down base in the smaller model.

(c) How will the total surface area of the model *compare* (this means an appropriate *ratio*) with the total surface area of the original? How will the surface areas of the circular bases compare?

(d) If both the model and the original are made of the same material, how will the mass of the model compare with the mass of the original?

(e) If the model and the original are *not* made of the same material, what would you have to know about the materials to be able to compare the masses, and how would you use this information?

(f) If the original statue and the model turned out to have the same mass, what would you conclude about the materials making up the two objects? (Give a *numerical* answer comparing relevant properties of the materials.)

1.7 As a first approximation, let us think of an older person O as being a larger scale model of a younger person Y, with lengths in all parts of the body increased by the same scale factor. Suppose the weight of O is 4.50 times the weight of Y.

(a) If the height of O is 6.0 ft, what must be the height of Y?

(b) How will the cross-sectional area A_O at some level in the legs of the older person compare with the cross-sectional area A_Y at the corresponding level in the legs of the younger?

(c) Will the compressional *stress* on the leg bones be the same in the two individuals? ("Stress" is the name for the force per unit area that the bone must support.) Why or why not? If not, for which individual will the stress be greater? By what numerical factor? It is because of the effect that emerges in the analysis you have just conducted that larger animals are, in reality in the world around us, *not* simply scaled-up models of smaller animals. Explain the reasoning behind this statement.

1.8 In explosions, the radius of a particular level of damage (say, the collapse of wooden buildings) varies as the cube root of the energy released in the explosion. (Energy release is frequently measured in terms of the equivalent number of pounds of TNT.) Suppose that in case A the energy release is equivalent to 200 lb of TNT while in case B the release is equivalent to 1200 lb.

(a) How will the radii for a given level of damage compare in cases A and B? Explain your reasoning.

(b) How will the *area* in case A, S_A, compare with the area in case B, S_B, for the same level of damage? Explain your reasoning.

1.9 A curve of radius R in a highway is banked at the optimum angle for cars traveling at a speed of 55 mi/h. By what factor must the radius of the curve be increased or decreased if the banking angle is to remain the same and still be the optimum value for trucks having a mass of 10,000 kg traveling at a speed of 40 mi/h? (This is a problem in ratio reasoning. Go to the equation that was derived for optimum angle of banking, establish the functional relation relevant to this problem, and calculate the scaling factor called for. Explain your reasoning, being careful to indicate what role is played by the mass and what role is played by the radius.)

1.10 The planet Saturn has a large moon called Titan. Titan has a mass 1.85 times the mass of our moon. Saturn itself has a mass 95 times that of the earth. Our moon has a mass 0.0123 times that of the earth. The distance between the centers of earth and moon is 240,000 miles, and the distance between centers of Saturn and Titan is 760,000 miles. Referring to the law of gravitation and explaining your reasoning in terms of scaling ratios *only* (without substitution in the formula), calculate the ratio of the centripetal force F_{TS} exerted on Titan by Saturn with the centripetal force F_{ME} exerted on the moon by the earth. (Give your argument in terms of whether F_{TS} will be larger or smaller than F_{ME}, and in what ratio, as prescribed by the gravitation law.)

1.11 The center of the moon executes a nearly circular orbit, with a radius of about 240,000 miles, around the center of the earth. Suppose the circumference of this orbit were increased by a length of 63 ft. (One mile contains 5280 ft.) By what amount would the earth–moon distance change? (Hint: To make the simplest and most efficient calculation, avoiding foolish complications with irrelevant numbers, sketch a graph of circumference versus radius for circles and locate the relevant circumference change, the corresponding radius change, and the relation between the two changes on your graph before plunging into calculations. If you find yourself working with huge numbers and converting between miles and feet, you are on a wrong track.)

Note to the instructor: In question 1.12, "density" is deliberately *not* mentioned. Use of this word directs many students to the formula for density and diverts them from the arithmetical reasoning.

1.12 You have a block of wood with a total mass of 540 kg. This type of wood has 0.85 g in each cubic centimeter. Suppose you were to add 38 g of wood to the block. By how much would you increase the volume of the entire block? Do *not* substitute into a formula; explain the relevant arithmetical reasoning in your own words. (Hint: Be sure to think about *change* in volume rather than entire volume; look at the relation between the lines representing volume changes and corresponding mass changes on the graph of total mass versus total volume, and avoid making useless and irrelevant calculations.)

1.13 It is an empirical fact that the power output required of the engines of a boat or ship varies approximately as the cube of the speed; i.e., if you wish to double the speed of the vessel, you must increase the power output by a factor of 8.

(a) Consider a boat with a mass of 2000 kg moving with initial speed v_i. The captain increases the power output of the engines by a factor of 2.6. By what factor does he increase the speed of the boat? Explain your reasoning.

(b) By what factor does he increase the kinetic energy of the boat? Explain your reasoning.

1.14 Suppose a photographer has established the correct exposure time for a particular situation as determined by the brightness of the scene, the film being used, the lens, and the diameter of the lens opening. He now wishes to shift from his 50 mm focal length lens to a 120 mm lens and to reduce the exposure time by a factor of 3.6 while the light conditions and the film remain unchanged. Must he increase or decrease the diameter of the lens opening? By what numerical factor? Explain your reasoning.

1.15 It is established that oxygen atoms have 1.33 times the mass of carbon atoms. Suppose we have 100 g of oxygen and we want to weigh out an amount of carbon that has the same number of atoms as our sample of oxygen. How many grams of carbon should we weigh out? How would the number of atoms in 100 g of carbon compare with the number in 100 g of oxygen? Explain your reasoning in your own words. (Note that at this stage, we have no idea how many atoms of oxygen are actually present in 100 g of the gas, and we have no need for this information.)

1.16 Consider the case of interaction between two point charges A and B exerting a force of 0.0126 N on each other. Suppose that the magnitude of charge A is increased by a factor of 3.25, the magnitude of charge B is decreased to 0.873 its initial value, and the spacing between the charges is increased by a factor of 1.18.

(a) Referring to Coulomb's law to justify the ratio reasoning that is appropriate, calculate the final force acting on charge A by multiplying the initial force (0.0126 N) by numerical factors that increase or decrease the force in accordance with the changes that are described. Do *not* substitute into the formula, but indicate the physical justification for each ratio applied. Put your calculation as a sequence along the following straight line:

0.0126 N ✕

1.17 Two point charges q_A and q_B are separated by a distance of 1.80 cm and interact with a force of 0.0035 N. Charge q_A is increased by a factor of 8.2 and the separation is increased to 4.30 cm. By what factor must charge q_B be changed if the force of interaction is to end up at a value of 0.0045 N? (Solve the problem directly by ratio reasoning without formal substitution into Coulomb's law.) Explain your reasoning.

1.18 Two point charges A and B attract each other with a particular value of force. Suppose that the magnitude of charge A is increased by a factor of 5.3 while that of B is decreased by a factor of 2.4. By what factor must the spacing between the charges be changed to keep the force of interaction unchanged from its initial value? Be sure to explain your ratio reasoning and to indicate whether the spacing is to be increased or decreased.

1.19 In the electrolysis of water, the passage of an electrical charge of 96,500 C liberates 1.008 g of hydrogen. From its chemical properties, we know that the hydrogen ion is associated with one corpuscle of electrical charge, i.e., is represented by the symbol H^+.

(a) Calculate the charge-to-mass ratio for H^+ in coulombs per kilogram.

It is known that the nitrogen atom (N) has a relative mass of 14.0 on the scale in which hydrogen has the relative mass 1.008. It is also known that ordinary nitrogen gas is diatomic; i.e., it has the molecular formula N_2.

(b) Suppose you are looking for doubly ionized nitrogen molecules N_2^{2+} in a beam of positive ions. Reasoning in terms of available ratios, calculate the expected charge-to-mass ratio of N_2^{2+} in coulombs per kilogram, starting with the value for H^+ and using ratio reasoning. Explain your reasoning.

1.20 Consider the following names for different functional relationships: (1) "direct proportionality," (2) "linear relation but not a direct proportionality," (3) "nonlinear relation," (4) "inverse proportionality," (5) "inverse square relation."

(a) Sketch a graph illustrating each relationship on a set of coordinate (x and y) axes.

(b) Write an algebraic equation in terms of y and x corresponding to each illustration.

(c) Illustrate each type of relation by describing a physical situation in which you have encountered it.

1.21 Consider the following graphs:

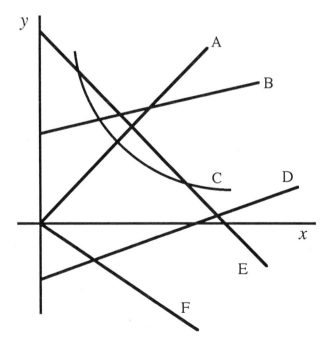

(a) Which, if any, of the graphs represent direct proportions?

(b) Which, if any, of the graphs represent linear relations that are *not* direct proportions?

(c) Which, if any, of the graphs represent nonlinear relations?

(d) Which, if any, of the graphs might represent inverse proportions?

1.22 Consider the following argument between two individuals who are comparing two bodies (#1 and #2) of possibly different composition by comparing their densities. Person A argues that since the mass of #2 is twice that of #1, the density of #2 must be twice that of #1 because density is directly proportional to mass. Person B, on the other hand, argues that since the volume of #2 is twice that of #1, the density of #2 must be half that of #1 because density is inversely proportional to volume.

(a) Assuming the information about masses and volumes is correct, what is the actual relation between the two densities? Explain your reasoning.

(b) Describe in your own words just where the fallacy in the arguments presented by A and B lies.

1.23 Centripetal acceleration of an object in uniform circular motion can be expressed in two alternative forms:

$$a_c = v^2/r \quad \text{or} \quad a_c = r\omega^2$$

where r denotes the radius of the circle, v the magnitude of the tangential velocity of the object, and ω its angular velocity. Since $v = r\omega$, it is clear that the two expressions are equivalent.

Using the first expression, person A argues that since centripetal acceleration is inversely proportional to r, centripetal acceleration of the object must be larger in smaller circles than in larger ones. Person B disagrees and, appealing to the second expression, argues that since centripetal acceleration is directly proportional to r, centripetal acceleration of the object must be smaller in smaller circles.

(a) Starting with the definitions of v and ω and with the help of a simple diagram, explain where the relation $v = r\omega$ comes from.

(b) Identify the source of the disagreement between A and B in the foregoing arguments, and alter each statement to make it correct and eliminate the inconsistency between the two.

1.24 Consider the following two equations :

$$(1) \ y = 3x - 5 \quad \text{and} \quad (2) \ \bar{v} = \Delta s / \Delta t$$

where the latter refers to the concept of average velocity in kinematics.

(a) Describe in your own words the difference between statements (1) and (2) as far as their origin, nature, and logical content are concerned. [Hint: How did (2) originate? Why would it be more appropriate in (2) to use the symbol $\equiv$ (meaning "defined as" or "identical to") rather than the ordinary $=$ sign?]

Consider two additional equations:

$$(3) \ (x + y)^2 = x^2 + 2xy + y^2 \quad \text{and} \quad (4) \ \cos 60° = 0.500$$

(b) What differences do you discern among the meanings of the equals signs in equations (1), (3), and (4)? Is (3) a functional relation in the same sense as (1)? Why or why not? Is (4) a functional relation? Describe the differences in detail in your own words, with appropriate illustrations or examples.

Consider the two equations:

$$(5) \ 2y = 5x \quad \text{and} \quad (6) \ 2 \, \text{in.} = 5 \, \text{cm}$$

where (5) is an ordinary algebraic relation and (6) is the relation between the different measures of length (i.e., inches and centimeters).

(c) Indicate the ways in which the statements (5) and (6) are profoundly different in meaning even though the same symbol ($=$) is (somewhat misleadingly) used in both. [Hint: Is 2 times the number of inches equal to 5 times the number of centimeters? (That is, are "in." and "cm" in (6) similar in meaning to the symbols y and x in (5)?)]

(d) What about equations such as $F_{net} = ma$ or $F = Gm_1 m_2 / r^2$? Does the equals sign have the same meaning it does in the preceding examples? (Do not forget the extended "stories" and definitions that go with generating the meaning of these two relations.)

(e) Does the equals sign always have exactly the same meaning wherever it happens to arise in your textbooks? Explain to a fellow student who is having trouble seeing the point of this entire question what it tells you about the caution you should exercise in connection with casual textbook usage of the same symbol ($=$) in very different contexts and physical situations.

1.25 You have probably seen a film in which a flower, starting as a bud, opens up before your eyes. This speeding-up effect is accomplished by "time lapse photography," in which the frames are snapshots, taken with a carefully chosen time interval between them and not as a continuous "motion picture." In this problem you are asked to make your own estimates and judgments as to the time interval between frames that will yield the desired speeding-up effect. (Estimating is not a matter of wild guesswork without underpinnings. It is a matter of careful reasoning with meaningful, if not highly precise, values.)

(a) Suppose you want to present on the screen the complete opening of a flower (daffodil? rhododendron? camellia? anything you like) from the initial bud to the fully opened flower. How long do you want the entire sequence to last when it is finally projected?

(b) In films, the illusion of motion is produced by flashing successive pictures on to the screen at the rate of about 20 frames per second. How many pictures will you need to present your entire speeded-up version of the opening of the flower? About how long does it take for the flower to open? (You must make your own reasonable estimates of relevant numerical values.)

(c) Time lapse photography is carried out by focussing on a given object a movie camera equipped with an automatic triggering device that permits successive snapshots to be taken on successive frames of film at intervals of seconds, minutes, hours—whatever is desired—instead of running the film continuously at high speed. To obtain your movie of the flower, what time lapse (i.e., what time interval between successive frames) will you want to set? Carry out the calculation and explain your reasoning.

(d) Now explain to a novice, invoking numerical data and not just words, why you would *not* want to photograph the sequence at normal camera speed and then run it very much faster to show the flowering in "slow motion."

1.26 The speed of light is known to be about 186,000 mi/s. Use denary (power of ten) notation in making the following calculations, and explain your reasoning throughout. Do not use formulas in which to substitute. Present your solutions in terms of simple arithmetic and purely arithmetical reasoning. Do not give numerical answers to any more than the justifiable number of significant figures.

(a) Light from the sun requires about 8 minutes to travel from the sun to the earth. How far from earth is the sun (in miles)?

(b) The average distance from the earth to the moon is about 240,000 miles. How long does it take a flash of laser light to travel from the earth to a target on the moon?

(c) Read the following question regarding mass and volume for understanding, but do not answer it.

Metallic copper contains a mass of 8.9 g in each cubic centimeter. The volume of a copper bar is known to be 50 cm^3. What is the mass of the bar?

Here is the question to be answered: Is the reasoning to be used in the question regarding the copper more like that used in part (a) or in part (b)? Explain your answer.

(d) Make up a problem about the metallic copper in which the reasoning to be used would be analogous to whichever [either (a) or (b)] did not apply to (c).

1.27 It is an experimental fact that the total electrical resistance of any metallic wire is directly proportional to its length and inversely proportional to its cross-sectional area (in other words, inversely proportional to the square of the diameter). The proportionality constant is, of course, different for each different metal and is thus a property of any given metal.

(a) In your own words, argue that these experimental facts indicate that in the metallic wire, electric charge must be moving throughout the entire *cross section* of the wire and is not, for example, confined to the surface. If the current were confined to the surface, what would you expect to have been the relation between resistance and wire diameter? Explain your reasoning.

(This is an example of how important it sometimes is to recognize what is *not* the case and contrast it with what *is* the case. The resistance might conceivably have been inversely proportional to the diameter, but, in fact, it is not. It is an experimental fact, however, that if current in the wire is alternated at extremely high frequency, the motion of charge *is* confined to the surface. This so-called skin effect does not play a role in circuits with which we are concerned in introductory physics.)

(b) Suppose a solid rod of metal has a radius a, a length L, and a resistance R. A hollow rod is now made of the same material with the same length L and the same outer radius a, but its inner radius is half the outer radius. How will the resistance of the hollow rod compare with that of the solid rod? Obtain the numerical value of the ratio and explain your reasoning.

1.28 A point on the earth's equator has a tangential velocity v_{rot} (relative to the fixed stars) by virtue of rotation about its own axis. The earth also has a tangential velocity v_{rev} in its revolution about the sun. The radius of the earth's solar orbit is about 23,400 times as large as the earth's diameter.

(a) Using your everyday knowledge of relevant time intervals, compare the two tangential velocities. Which one is larger? How much larger? Use ratio reasoning without substitution in formulas and explain each step.

(b) How do the tangential velocities of rotation at points at latitude 45°N or S and at the North Pole or South Pole compare with the tangential velocity of revolution around the sun? Draw an appropriate diagram and explain your reasoning.

(c) [For students who might have studied the Michelson–Morley experiment: What relevance do these comparisons have to the Michelson–Morley experiment?]

1.29 A car starts from rest at position $s = 0$ and accelerates uniformly along a straight road in the positive s direction. At position s_1 it has a velocity v_1. What will be its velocity at positions $2s_1$ and $3s_1$? Express your result in terms of v_1 multiplied by a numerical factor and explain your reasoning. (Do your reasoning in terms of the ratios called for by the applicable functional relation, not by substitution in a formula.)

1.30 A ball rolls down an inclined plane of length L starting from rest at the top. Where along the plane would it have half the velocity it reaches at the bottom? Explain your reasoning.

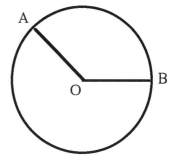

1.31 Along with the accompanying figure you are given a ruler and a piece of thread.

(a) Describe in your own words what you would do to obtain the magnitude of angle AOB in radians. (In effect, you are being asked to give an operational definition of radian measure.)

(b) Explain in your own words why and under what circumstances it is desirable to utilize radians rather than degrees in measuring angles. What is inadequate about degrees?

CHAPTER 2

Kinematics

2.1 In cases of rectilinear motion, we give the name "average velocity" and the symbol $\bar{v}$ to the quantity $\Delta s/\Delta t$, where s represents position numbers along the straight line in question and t represents clock readings at corresponding values of s. $\bar{v}$ is a single number characterizing the motion during the time interval Δt regardless of how complex and variable the velocity history may have been. It is a vector quantity in the sense that it is accompanied by plus or minus signs depending on the algebraic sign of Δs. The symbol $|\bar{v}|$ would be described as the "magnitude of the average velocity for the given time interval," meaning the size of this quantity regardless of its algebraic sign (direction).

> (a) Consider the quantity that would be described as the "average magnitude of the instantaneous velocity" and would be represented by the symbol $\overline{|v|}$. How would you go about calculating $\overline{|v|}$? Describe the process in detail.

> (b) Describe some motions in which the two quantities $|\bar{v}|$ and $\overline{|v|}$ would come out equal. Describe some motions for which they would *not* be equal. Include in your discussion the case in which the body, at the end to the interval Δt, returns to the same position it occupied at the beginning.

2.2 If we are dealing with a situation in which the instantaneous velocity v changes significantly over the given time interval Δt, what is wrong with saying $\Delta s = v\Delta t$? What is the only value of velocity that will make this equation correct under these circumstances? Explain your reasoning.

2.3 Consider making the calculation $\Delta t / \Delta s$ instead of $\Delta s / \Delta t$. Is this just a piece of foolishness, or does the number so obtained have a reasonable and intelligible physical meaning?

(a) Interpret $\Delta t / \Delta s$ in words, noting that the interpretation, in words, of $\Delta s / \Delta t$ is "the number of meters change of position taking place in *one* second."

(b) Execute with movements of your hand (1) a motion for which $\Delta t / \Delta s$ would have a very high value and (2) a motion for which it would have a very low value, and contrast your actions with those you would execute if illustrating $\Delta s / \Delta t$.

(c) Invent a good, descriptive name to give the quantity defined by $\Delta t / \Delta s$. (Geophysicists actually do have a descriptive name for this quantity; see if you can guess what it is in the light of the character of the motion when $\Delta t / \Delta s$ is very large.)

2.4 In this question, we shall do some qualitative reasoning concerning the effect of air resistance on the motion of a ball thrown vertically upward with initial velocity v_{oy}, rising to its maximum height, and then falling back to the level from which it was thrown. (The velocity of the thrown ball is great enough for air resistance to play a significant role but not so great that the ball approaches terminal velocity at any time during its flight.) We raise the following question: How does the time interval $(\Delta t)_{up}$ for rise to maximum height compare with the time interval $(\Delta t)_{down}$ for falling from maximum height back to the starting level? Record your initial intuitive response to this question: Are the two time intervals equal or unequal? If unequal, which is greater? Explain the basis for your intuitive response.

Now let us analyze the situation through a careful sequence of questions, making use of the following notation. Under conditions of negligible air resistance, the magnitude of the vertical acceleration is denoted by g. Let us denote the magnitude of the average acceleration during the upward motion in the presence of air resistance by a_{uy} and the magnitude of the average acceleration during the downward motion by a_{dy}.

(a) How would you expect the height of rise in the presence of significant air resistance to compare with the height in the case of negligible air resistance? Explain your reasoning by referring to the relevant kinematic equation.

(b) How would you expect the magnitude of a_{uy} to compare with g? Would it be greater than, equal to, or smaller than g? Explain your reasoning. Is your answer consistent with the answer you gave in part (a)? Explain.

(c) How would you expect the magnitude of a_{dy} to compare with g? Explain your reasoning.

(d) In the light of your answers in parts (b) and (c), how will the time interval $(\Delta t)_{up}$ for rising to maximum height compare with the time interval $(\Delta t)_{down}$ for falling from maximum height back to the starting level? Explain your reasoning. Is the result you have arrived at consistent with the one you anticipated intuitively at the start of the question? If not, reexamine what you have done, with the objective of achieving internal consistency. (Hint: It helps to keep in mind that the *distances* of rise and fall are the same.)

2.5 A person standing at the edge of the roof of a building throws ball A vertically upward with an initial velocity $+\left|v_{oy}\right|$ and throws a second ball B vertically downward with an initial velocity $-\left|v_{oy}\right|$ of the same magnitude. Both balls fall to the ground past the edge of the building.

(a) Under conditions of negligible air resistance, what is the velocity of ball A as it passes, on the way down, the level from which it was thrown? Explain your reasoning?

(b) Under conditions of negligible air resistance, how will the velocities of the two balls on striking the ground compare with each other? Explain your reasoning.

(c) Now let us visualize the same experiment performed in the presence of significant air resistance. How will the two velocities on striking the ground compare with each other? Will they be equal in magnitude or unequal? If unequal, which will be greater and why?

(d) Suppose that a third ball C is projected from the edge of the roof with the same vertical component of velocity $+\left|v_{oy}\right|$ but with a horizontal component of the same magnitude. In the case of negligible air resistance, how will the magnitude of the *total* velocity of C on striking the ground compare with that of A? (Draw a relevant diagram; find the numerical value of the ratio; explain your reasoning.)

2.6 You have an ordinary stopwatch such as that used in timing athletic events. Describe how you might take advantage of the relation $\Delta s = (1/2)g(\Delta t)^2$ and the known value of g to determine the height of a window above the ground (or the height of a bridge above a stream) by dropping an object from the upper location. Examine the accuracy to be expected under various circumstances: What trouble would you run into if the height is relatively small? How small is "relatively small"? About how large would the height have to be for you to obtain reasonably reliable values? What trouble do you begin to run into as the height becomes quite large?

2.7 Consider a car moving along a highway. Answer the following questions, giving an explanation of your answer in each case. Sketch at least one possible set of a versus t, v versus t, and s versus t graphs corresponding to your description in each case.

    (a) If the acceleration a of the car is zero, what are the possible values of velocity ?

    (b) If the car is moving, is it necessarily accelerating?

    (c) If the car is not accelerating, is it necessarily standing still?

    (d) Under what circumstances is the acceleration in the opposite direction to the velocity?

    (e) How might you drive a car so that the acceleration would go through zero (from positive to negative) while the velocity remains positive?

    (f) How might you drive a car so that the velocity would go through zero (from negative to positive) while the acceleration remains positive?

2.8 Suppose you are driving a car and are accelerating, increasing your speed in the positive direction. You now relax slowly on the gas pedal, decreasing the *magnitude* of your acceleration (you do *not* use the brake).

    (a) Are you increasing or decreasing your speed as the *magnitude* of the acceleration decreases? Explain your reasoning.

    (b) Sketch a versus t and v versus t graphs for the situation under consideration and make sure that your graphs are consistent with the verbal description you gave in part (a).

    (c) Sketch corresponding graphs for the same sequence except for initial motion in the negative direction.

2.9 A numerical value of acceleration can be interpreted as telling us "how fast the velocity of a body is changing." Starting with the *definition* of acceleration, explain why this is a legitimate statement. Now consider the following statement: "A numerical value of acceleration, given alone, tells us nothing about how fast the object in question is moving." Is this statement correct or incorrect? Explain your answer carefully by referring to the definitions of both velocity and acceleration.

2.10 Consider the situation sketched in the figure: at clock reading $t = 0$, observer A at the edge of the cliff throws a ball vertically upward with an initial velocity of 30 m/s. Observer B is located in the helicopter, which started at the base of the cliff and is rising vertically at the constant velocity of 30 m/s. At the same clock reading $t = 0$ at which A throws his ball, B is passing A and releases a ball from his window. (B simply lets go the ball without any throwing action.) B continues upward in the helicopter with no change in the upward velocity of 30 m/s.

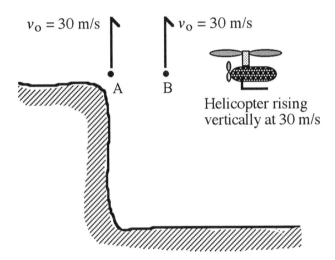

(a) Describe how observer A perceives the motion of the two balls relative to his position on the edge of the cliff after the instant $t = 0$. That is, what does each ball do relative to this observer? How does A describe the velocity as varying? How does A describe the acceleration? Cover the entire sequence between $t = 0$ and the instant the balls finally land at the base of the cliff.

(b) Describe how observer B perceives the motion of the two balls, not relative to the ground but relative to B's frame of reference in the rising helicopter. In particular, what is each ball doing relative to B at the instant A claims that the ball has reached the top of its flight? How does B describe the velocity as varying? How does B describe the acceleration?

2.11 The following velocity versus clock reading histories describe the rectilinear motion of six particles that started out from position s_0 at $t = 0$ s. Circle the correct answers for each of the following questions.

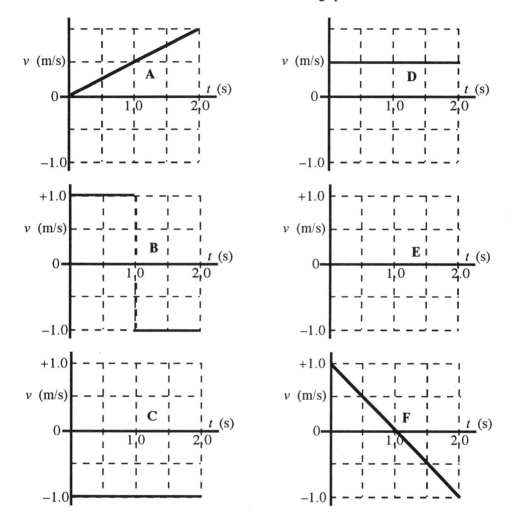

(a) Which particle (or particles) have returned to position s_0 at the clock reading $t = 2.0$ s?

A   B   C   D   E   F   NONE

(b) Which particle (or particles) spend at least some time moving in the negative s direction?

A   B   C   D   E   F   NONE

(c) Which particle (or particles) move at uniform, nonzero acceleration?

A   B   C   D   E   F   NONE

(d) Which particle (or particles) started in the negative s direction and then reversed the direction of motion, traveling back in the positive s direction?

A   B   C   D   E   F   NONE

(e) Which particle is farthest from position s_0 at clock reading $t = 2.0$ s?

A   B   C   D   E   F   NONE

(f) Which particles are the same distance from s_0 at $t = 2.0$ s?

A   B   C   D   E   F   NONE

(g) Which particle (or particles) exhibited nonzero acceleration during the given period?

A   B   C   D   E   F   NONE

(h) Which particle (or particles) kept moving in the same direction throughout the given period?

A   B   C   D   E   F   NONE

(i) Which particle (or particles) exhibited negative acceleration over some interval?

A   B   C   D   E   F   NONE

(j) Which particle (or particles) stood still for some time at some point in the history?

A   B   C   D   E   F   NONE

(k) Which particle (or particles) move with nonuniform, increasing acceleration?

A   B   C   D   E   F   NONE

2.12 The diagram shows the position s versus clock reading t history of the motion of a car starting at $t = 0$.

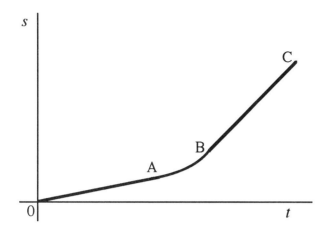

Describe in your own words what you would see the speedometer needle doing during the various portions of the history: from 0 to A, from A to B, from B to C. Does any acceleration take place? If so, over what interval?

2.13 Consider the three different histories of velocity v versus clock reading t shown in the figure.

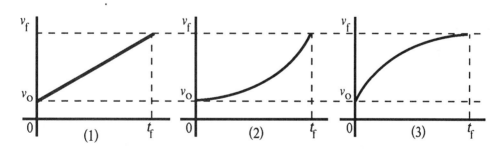

(a) In case 1, argue that the average velocity $\bar{v}$ over the time interval between $t = 0$ and $t = t_f$ must be equal to $(v_0 + v_f)/2$. Explain your reasoning carefully, making use of the fact that the history is a straight line.

(b) In cases 2 and 3, is $\bar{v}$ also equal to $(v_0 + v_f)/2$? Why or why not? If not, how does $\bar{v}$ compare with $(v_0 + v_f)/2$ in each instance? Is it greater or smaller? Explain your reasoning.

2.14 The graphs show the position–clock reading histories of the simultaneous motions of two cars A and B in parallel lanes along a straight highway in two different occurrences.

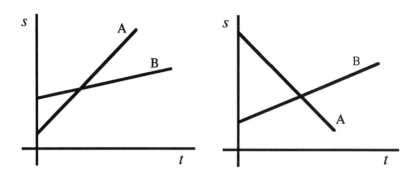

(a) Are there any instants at which the two cars have the same velocity? If so, mark the instant (or instants) along the t-axis with the symbol t_{eq} and explain how you arrived at your conclusion. If not, explain how you arrived at that conclusion.

(b) Are there any instants at which the two cars are passing each other, i.e., are at the same distance from the origin of position s? If so, mark the instant or instants with the symbol t_p. Explain your reasoning. In each case, which car is going faster at the instant of passing? Explain your reasoning.

(c) At the instant the cars are located at the same positions s, have they traveled the same or different distances from their own initial positions at $t = 0$? If the distances are unequal, indicate which is larger. Explain your reasoning.

(d) Suppose the *speed* of car B is increased somewhat in each case. What will happen to the clock reading at which the two cars will be located at the same position? What will happen to the distance each car will have travelled from its $t = 0$ position? Explain your reasoning by altering the diagrams to show what happens in each case.

2.15 Suppose that you are driving in car A along a straight road at 10 mi/h. A friend in car B is driving at a constant velocity of 60 mi/h (88 ft/s). At clock reading $t = 0$ and position $x = 0$, car B passes you and continues without change in velocity. You, however, in car A, step on the accelerator and maintain a constant *acceleration* of 5.0 ft/(s)(s).

(a) Sketch a *qualitative* (do not try to plot numerically) graph of the x versus t history of the motion of the two cars.

(b) On the basis of the diagram you have sketched, infer whether or not you will ever overtake car B. Explain your reasoning.

(c) If, in part (b), you concluded that you *will* overtake B, calculate the clock reading at which the overtaking will occur and calculate the velocity (in both mi/h and ft/s) you will have attained at the instant of overtaking. Be sure to examine and interpret your results to determine whether or not the plan for overtaking is realistic.

2.16 A ball is fired off the edge of a table with a horizontal velocity v_x and lands on the floor.

(a) On the diagram, sketch a possible trajectory (the path followed by the ball) from the edge of the table to the floor.

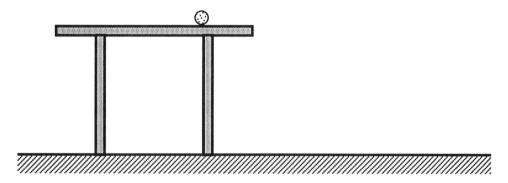

(b) Now, on the same diagram, sketch two other trajectories, one for a smaller value of v_x and one for a somewhat larger value of v_x. Sketch a fourth trajectory for an extremely large value of v_x. Label each of the trajectories with the comparative sizes of v_x.

(c) We say that the shape of such trajectories is parabolic in the ideal case in which air friction is negligible. What is a "parabola"? That is, how is this kind of curve defined? How do we know that the shape of the trajectory is parabolic?

(d) For the drop of the ball from its original level at the height of the table to be virtually unobservable to the naked eye at the wall of the room some reasonable distance from the table, what would have to be the numerical magnitude of v_x? (You will have to choose your own reasonable values for the drop and for the distance to the wall.) What effect would you expect the ball to have on the wall under these circumstances? (Justify your answer.)

(e) Return to part (a) and sketch another trajectory: that of another ball having a very much larger mass than that of the first ball but exactly the same initial velocity v_x. Explain your reasoning.

2.17 The figure shows position versus clock reading histories of rectilinear motions of two balls A and B rolling on parallel tracks.

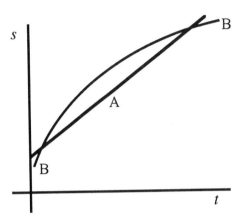

(a) Mark with the symbol t_p along the t-axis on the diagram any instant or instants at which one ball is passing the other.

(b) Which ball, A or B, is moving faster at any clock reading t_p?

(c) Mark with the symbol t_{eq} along the t-axis any instant or instants at which the two balls have the same velocity.

(d) Circle the correct statement from among the following: Over the period of time shown in the diagram, ball B is

speeding up all the time.

slowing down all the time.

speeding up part of the time and slowing down part of the time.

neither speeding up nor slowing down.

(e) Over the time interval between the passing points, does ball B travel a greater distance, a smaller distance, or the same distance as ball A?

2.18 The figure represents a flash (or stroboscopic) photograph looking down on two balls rolling parallel to a position scale on a level table top. The numbers show the clock readings corresponding to each ball position.

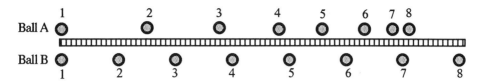

(a) At approximately what clock reading (or readings) do the two balls have very nearly the same speed?

(b) At approximately what clock reading (or readings) does one ball pass the other? In each instance you cite, indicate which ball, A or B, is doing the overtaking.

(c) Sketch position versus clock reading graphs for each of the two balls on the same set of axes, showing clearly how the two motions are related. Be sure to check whether your answers in parts (a) and (b) are consistent with your graphs. Explain your reasoning.

2.19 In the figure, we are looking down on a level table top. Assume that a flash (or stroboscopic) photograph has been taken of two balls A and B rolling parallel to a position scale on the table. Corresponding clock readings are shown at each image.

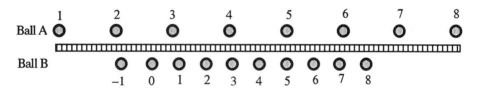

(a) At what clock readings, if any, do the two balls have very nearly the same speed?

(b) At what instant or instants, if any, is ball B overtaking and passing ball A?

(c) Sketch position versus clock reading graphs for each of the two balls on the same set of axes, showing clearly how the two motions are related. Check whether your answers to parts (a) and (b) are consistent with your graphs, and explain your reasoning.

2.20 Two balls A and B are released from rest and roll down sloping sections of track as shown. The slopes of the two tracks are *not* necessarily the same as those in the diagram. At the foot of each slope, the balls roll on to level sections of track along which they continue at uniform velocity. Certain measured times and distances are shown.

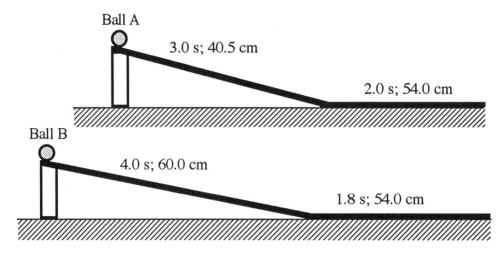

Ball A

3.0 s; 40.5 cm

2.0 s; 54.0 cm

Ball B

4.0 s; 60.0 cm

1.8 s; 54.0 cm

(a) According to the information given, which ball has the greater acceleration on its sloping section of track? Base your analysis *directly* on the fundamental definition of acceleration. No credit will be given if you use derived kinematic relations. Show all your calculations and explain your reasoning.

2.21 A pendulum bob, released from rest at position 1, can swing to position 5.

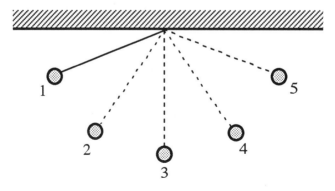

1

5

2

4

3

(a) Suppose that as the bob swings, the string is suddenly cut at position 2, or position 3, or position 4, or position 5. In each instance, sketch the trajectory the bob would follow if the string were cut at the instant the moving bob reached that position, and explain your reasoning.

2.22 The table shows histories of instantaneous velocity versus clock readings for two different cases in an accelerating car.

Clock reading t (s)	Case I		Case II	
	Speedometer reading v (mi/hr)	Average acceleration $\bar{a}$ [mi/(hr)(s)]	Speedometer reading v (mi/hr)	Average acceleration $\bar{a}$ [mi/(hr)(s)]
0.0	+6.0		+20.0	
1.0	+4.0		+10.0	
2.0	+2.0		+5.0	
3.0	0.0		+1.0	
4.0	−2.0		0.0	
5.0	−4.0		0.0	

Clock and Speedometer Readings (Instantaneous Velocities) in an Accelerating Car

(a) In the blank columns, enter values of average acceleration $\bar{a}$ corresponding to each time interval for case I and case II.

(b) Sketch (do not try to plot) a versus t, v versus t, and position s versus t diagrams for each of the two cases.

(c) Describe how, if your were driving, you might actually make a car execute (approximately) each one of the two motions.

2.23 You are called upon to explain to a fellow student how it comes about that the change in position Δs for an object in rectilinear motion during a time interval Δt might carry either a positive or a negative sign.

(a) Present your explanation, making use of appropriate sketches or diagrams and connecting your description with some easily visualizable situation, such as moving in a car or watching a cart on the table.

(b) Making use of the explanation you have given in part (a) and the definition of average velocity, explain to a fellow student how it comes about that velocities might be either positive or negative. Explain how the algebraic signs must be interpreted, connecting your explanations with the diagrams you utilized in part (a).

(c) Making use of the definition of acceleration and the explanations you gave in parts (a) and (b), explain to a fellow student how it comes about that acceleration values might come out either positive or negative. Explain how the algebraic signs must be interpreted, connecting your explanations with the experiences and diagrams you utilized earlier.

2.24 Consider the following data on the rectilinear motion of a car that starts from rest at clock reading $t = 0$ and position $x = 0$. At clock reading $t = 5.0\,s$, it is observed to be at position $x = +40.0\,m$ and to have an instantaneous velocity of $+11.0$ m/s.

(a) Examine the interconnections among the given data carefully. Was the acceleration of the car uniform? Explain your reasoning.

(b) Are the kinematic equations such as $v = v_o + at$ and $x = (1/2)at^2$ applicable throughout the history of the motion? Why or why not?

(c) Sketch the shape of the v versus t graph that is implied by the data: Is the graph straight or curved? If curved, is it concave up or concave down? Explain your reasoning.

2.25 The figure shows a schematic position s versus clock reading t history of the rectilinear motion of a body.

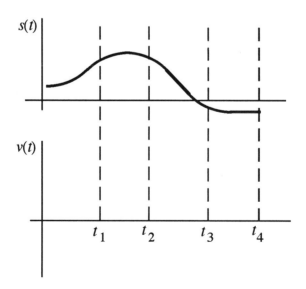

(a) On the $v(t)$ versus t coordinates, sketch the corresponding velocity versus clock reading history.

(b) The number $\int_{t_1}^{t_3} v(t)\,dt$ has a geometrical interpretation on the $v(t)$ versus t diagram. Indicate on this diagram what this interpretation is.

(c) The number $\int_{t_1}^{t_3} v(t)\,dt$ also has a geometrical interpretation on the $s(t)$ versus t diagram. Indicate on this diagram what this interpretation is.

2.26 A cart rolls down an inclined plane starting from rest at the top. It is halfway down after a time interval Δt.

(a) Will the time interval to get the rest of the way to the bottom be equal to, greater than, or less than Δt? Explain your reasoning.

2.27 A car starts from rest at position $s = 0$ and moves in the positive s direction. At position s_1 it is observed to have a velocity v_1. Some time later, at position $2s_1$, it is observed to have a velocity $2v_1$. At position $3s_1$, it has a velocity $3v_1$, etc.

(a) Is the car accelerating? If it is accelerating, is the acceleration uniform, increasing, or decreasing? Explain your reasoning.

2.28 Over a certain time interval Δt, a car moves in such a way that the magnitude of its instantaneous velocity is never *smaller* than the magnitude of the average velocity.

(a) Sketch a possible s versus t graph and comment on whether the car could have been accelerating during the given interval. Examine, in the same way, the case in which the magnitude of the instantaneous velocity is never *larger* than the magnitude of the average velocity.

(b) Sketch s versus t graphs for two cases: one in which the car is speeding up uniformly and the other in which it is slowing down uniformly over the same time interval Δt, making the average velocity the same in both graphs. Making use of the graphs, comment on how instantaneous velocities must compare with the average velocity at various instants during each of the two histories: Must there always be instantaneous velocities both greater and smaller than the average? Why or why not? Must there always be an instantaneous velocity that is equal to the average velocity? Why or why not?

(c) For those who have studied calculus: Can you connect the graphical and intuitive observations you make in parts (a) and (b) with any general theorems you developed in the calculus?

2.29 Consider a case in which two cars, A and B, on a straight road are located one behind the other as shown.

(a) Suppose car A starts from rest with uniform acceleration a at instant $t = 0$, moving in the positive direction (to the right) along the road. Car B starts somewhat later, at instant $t = t_1$, with exactly the same acceleration. What will happen to the spacing between the two cars as time goes by (while they are still accelerating)? Will they remain the same distance apart? Will the spacing keep decreasing? Will the spacing keep increasing? Explain your reasoning and support it both with relevant diagrams and with an algebraic analysis.

(b) Suppose we invert the situation as follows: car A has an instantaneous velocity v_o to the right at $t = 0$. Car B, somewhat behind car A, has the same instantaneous velocity at a somewhat later instant $t = t_1$. Both cars *slow down* uniformly with the same negative acceleration a. What will happen to the spacing between the two cars as time goes by (while they are still moving toward the right)? Will they remain the same distance apart? Will the spacing keep decreasing? Will the spacing keep increasing? Explain your reasoning and support it both with relevant diagrams and with an algebraic analysis.

2.30 A cart, released from rest at the top of an inclined plane, rolls down the plane, striking a spring at the bottom. It compresses the spring to the point at which its instantaneous velocity is zero and then, as the spring expands, is projected back up the plane, returning to the point at which it was released.

(a) Sketch s versus t and v versus t diagrams for the motion that has been described. Be sure to label the points at which the cart makes contact with the spring, has zero instantaneous velocity, and breaks contact with the spring.

2.31 The diagram shows a velocity vector v positioned relative to a set of coordinate axes y and x.

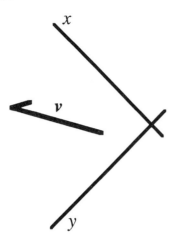

(a) Construct, on the diagram, the y and x *components* of the vector v. Explain your reasoning by describing how you use the definition of "component" in making your construction.

(b) Suppose the velocity vector is rotated counterclockwise through a small angle. Will the two components change in size? If so, describe how each one changes. Does it increase or decrease? Support your answers by making use of the diagram.

2.32 In these graphs, (a) represents the position versus clock reading history of the rectilinear motion of an object while (b) represents the velocity versus clock reading history of the motion.

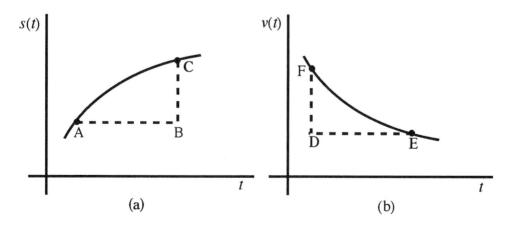

(a) What is the physical interpretation, if any, of the length of segment AB in diagram (a)?

(b) What is the physical interpretation, if any, of the length of segment BC in diagram (a)?

(c) What is the physical interpretation, if any, of the length of the diagonal segment AC (not drawn) in diagram (a)?

(d) What is the physical interpretation, if any, of the length of segment DE in diagram (b)?

(e) What is the physical interpretation, if any, of the length of segment DF in diagram (b)?

(f) What is the physical interpretation, if any, of the length of the diagonal segment FE (not drawn) in diagram (b)?

CHAPTER 3

Force and Dynamics

3.1 We have established that under the idealization of negligible air resistance, the trajectory of a projectile is a parabola, as that sketched here. Let us suppose for the moment that this is the trajectory that would have been followed, in the absence of air resistance, by a ball you have thrown.

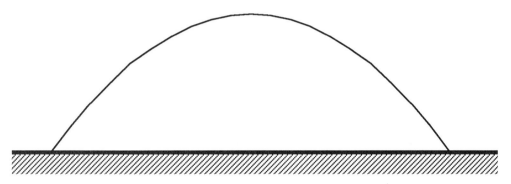

(a) Add to this sketch a trajectory you might expect the ball to follow in the presence of air resistance. You are not being asked for anything quantitative; the idea is to sketch *qualitatively* the change in shape you might expect. Explain your reasoning.

3.2 It is an observed fact that the force acting on any moving body due to air resistance increases as the velocity of the body increases. Make use of this fact, together with what you know about the motion of falling bodies, to present an argument predicting that raindrops of a fixed size, after accelerating downward for a time interval after their formation, will cease accelerating and attain a constant downward velocity (called "terminal velocity"). In making your argument, be sure to use several force diagrams showing what must be happening to the forces acting on the raindrop during the interval in which it is speeding up.

3.3 This figure is a snapshot looking down on a frictionless puck moving at uniform velocity v_o from left to right on a level air table. At the position shown, the puck is given a short, sharp hammer blow B in a direction perpendicular to that in which it is initially moving.

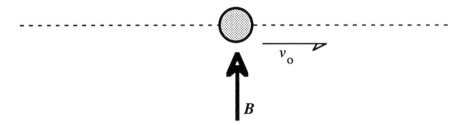

(a) Show on the figure a trajectory (or path) that the puck might follow on the table after the blow is delivered. Explain your reasoning.

(b) Will the final speed v_f of the puck (immediately after the blow) be equal to, greater than, or smaller than v_o? Explain your reasoning.

(c) How will the *velocity* of the puck on the frictionless surface behave as time goes by *after* the blow? That is, will either the magnitude or the direction of the velocity (or both) keep on changing? If so, how?

3.4 In this snapshot, we are looking down on a frictionless puck P which is moving counterclockwise in a circle on a level air table. (The puck is attached to a string, and the end of the string is fastened to a peg at point O.) At the instant the puck is in the position shown, the string is cut. Sketch the trajectory (or path) followed by the puck as it slides along the table after the string is cut. Explain your reasoning.

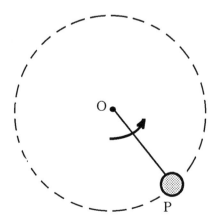

3.5 A glass tube formed into an incomplete circle lies on a level table as shown. The view is from above. A small marble M is blown into the open end of the tube at A. It swirls around the tube and emerges at the other open end at B. Sketch the trajectory (or path) that the marble will follow as it rolls along the table *after* leaving the tube. Explain your reasoning.

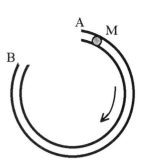

3.6 Consider a collision between a small car and a heavy truck. In such a collision, how does the size $F_{C \leftarrow T}$ of the force exerted on the car by the truck compare with the size $F_{T \leftarrow C}$ of the force exerted on the truck by the car instant by instant as the two are in contact? Is the first greater than, equal to, or smaller than the second? Explain your reasoning.

3.7 Consider the following statement: "In a tug-of-war, the force exerted by the *losing* side on the winning side must be *greater than* the force exerted by the winning side on the losing side." If you believe the statement to be correct, explain how the winning side manages to win under these circumstances. If you believe the statement to be incorrect, alter it so that it becomes correct and explain your reasoning. Finally, identify the force that makes it possible for the winning side to win.

3.8 Suppose you are pushing a cart along a level floor in the presence of frictional effects between the cart and the floor.

(a) While you are making the cart speed up, how does the size $F_{C \leftarrow Y}$ of the force you exert on the cart compare with the size $F_{Y \leftarrow C}$ of the force the cart exerts on you? Is the former greater than, equal to, or smaller than the latter? Explain your reasoning.

(b) While you are making the cart speed up, how does the size of the frictional force exerted by the floor on the cart compare with the size of the force the cart exerts on you? Explain your reasoning.

(c) Suppose you are now slowing the cart down from the speed you had imparted to it. How does the size $F_{C \leftarrow Y}$ of the force you now exert on the cart compare with the size $F_{Y \leftarrow C}$ of the force the cart exerts on you? Is the former greater than, equal to, or smaller than the latter? Explain your reasoning.

(d) While you are slowing the cart down, how might the size of the frictional force exerted by the floor on the cart compare with the size of the force you are exerting on the cart? Explain your reasoning.

3.9 Suppose you are throwing a ball vertically upward.

(a) While the ball is still in contact with your hand and you are accelerating it upward, how does the size $F_{B \leftarrow H}$ of the force your hand exerts on the ball compare with the weight W of the ball? Is the former greater than, equal to, or smaller than the latter? Explain your reasoning.

(b) While the ball is still in contact with your hand and you are accelerating it upward, how does the weight W of the ball compare with the size $F_{H \leftarrow B}$ of the force the ball exerts on your hand? Is the former greater than, equal to, or smaller than the latter? Explain your reasoning.

3.10 Suppose you are accelerating a very massive puck from rest on an air table by exerting a horizontal force. Can you accelerate the puck by exerting a force smaller than the weight of the puck? Explain your reasoning. How small a force will impart at least some acceleration to a very massive puck under perfectly frictionless circumstances? Do you think an ant, harnessed to the puck, would be able to get the puck moving under perfectly frictionless circumstances? Why or why not?

3.11 A block having a mass of 12.0 kg is held against the ceiling by a force $P = 160$ N acting at an angle of 75° to the horizontal as shown. It is known that block is in motion (sliding along the ceiling) and that the coefficient of kinetic friction is 0.20.

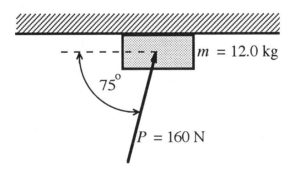

(a) Calculate the weight of the block. Explain your reasoning briefly.

(b) Sketch well-separated force diagrams of the block and the region of the ceiling in contact with the block. Describe each force in words and indicate the Third Law pairs.

(c) Calculate the normal force exerted by the ceiling on the block by formally applying Newton's second law to the block in the vertical direction.

(d) By formally applying Newton's second law to the block in the horizontal direction, determine whether the block is accelerating along the ceiling, and, if it is, calculate the numerical value of the acceleration. Explain your sequence of reasoning.

3.12 A pendulum bob, let go from rest at position 1, can swing over to position 5 as shown.

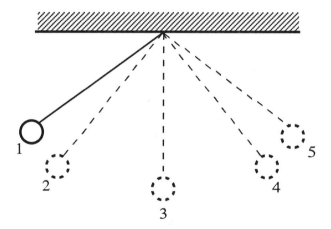

For *each* of the five indicated positions of the bob, draw three *separate* vector diagrams for the bob, as described in parts (a), (b), and (c).

(a) A diagram showing the instantaneous velocity vector for the bob.

(b) A diagram showing the forces acting on the bob.

(c) A diagram showing the instantaneous acceleration vector for the bob. (Keep in mind that the bob is *not* following a rectilinear path.)

(d) Finally, using vectors from parts (a) and (c), draw vector diagrams showing the *change* in velocity of the bob between positions 4 and 2 and the *change* in acceleration of the bob between positions 4 and 2. Explain how you draw each diagram.

3.13 Here we are looking down on a racetrack with straight sections and semicircular ends. A car is going around the track and maintaining constant speed.

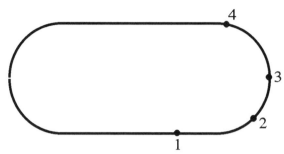

(a) For each one of the numbered positions: Draw diagrams showing the velocity vector for the car. On *another* set of diagrams, show the acceleration vector for the car. On a *third* set of diagrams draw the net horizontal force component acting on the car.

3.14 A string is attached to the ceiling at one end. A person pulls the string at the other end with a force F directed as shown. In the small diagram to the right, the person has sketched the horizontal and vertical components of the force exerted by the ceiling on the end of the string.

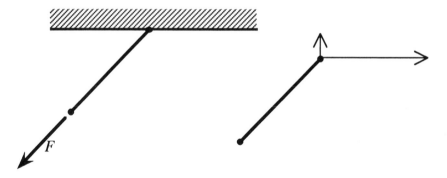

(a) Is the right hand sketch correct and reasonable? Explain your reasoning. (Note that such a sketch may be correct in some respects and incorrect in others.)

(b) Explain why the string in the diagram is straight. (Hint: Examine the forces acting on chunks of a curvy string, the accelerations that would be imparted to various chunks, and the circumstances under which the accelerations would become zero.) How would the situation change if the direction of the force were reversed? Explain your reasoning.

3.15 Consider a tank or pan containing water, as shown. When the water is first poured into the vessel, it swirls and sloshes around. This motion steadily dies down, however, indicating the existence of a nonzero acceleration and therefore of a net force opposing the motion of every swirling, sloshing parcel of water.

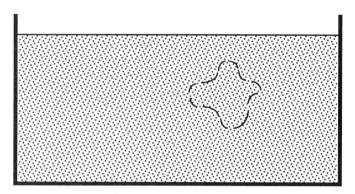

(a) How do you account for the dying down of the motion? Visualize the forces that might be acting on any arbitrarily shaped parcel of water. What effects can you visualize, even though they are not directly perceptible?

Once the motion in the vessel has died down, the accelerations are everywhere zero, and we infer that every parcel of water must be subjected either to no forces at all or to exactly balanced forces. Consider the particular, irregularly shaped parcel outlined in the figure. This parcel, like any other arbitrarily chosen parcel in the vessel, has its own weight; i.e., the gravitational force exerted by the earth is pulling it downward just as it pulls a book you hold in your hand.

(b) How do you account for the fact that the outlined parcel of water (and any other parcel of any shape whatsoever) does *not* fall or accelerate downward? What must be supplying the balancing upward force? Is the effect of the surrounding water distributed or concentrated? If distributed, over what region must it be distributed?

(c) Draw a force diagram for the outlined parcel, representing distributed forces by single arrows (as you would for a book resting on a table) rather than trying to show an actual distribution of forces.

Suppose you take a chunk of wood and submerge it in the water, holding the entire piece below the surface. Note that the wood has displaced (pushed out of the way) a parcel of water of exactly the same size and shape as the wood. As you have indicated in part (c), the original parcel of water that the wood has displaced was being held up by the surrounding water with a total force exactly equal to the weight of the parcel.

(d) Is there any reason to believe that this total upward force has changed simply because the original space is now occupied by the new object (the wood) instead of the original water? Explain your reasoning.

(e) Draw a force diagram for the chunk of wood while you are holding it submerged. Describe each force in words.

(f) How must the total upward force exerted by the surrounding water on the chunk of wood compare in magnitude with the weight of the chunk of wood: Is it greater, equal, or smaller? Explain your reasoning, noting the role that the relative densities of wood and water play in this context. Explain why you must exert a downward force on the chunk of wood to keep it submerged.

(g) What will happen to the chunk of wood (in terms of acceleration and change in velocity) after the instant you stop holding it? At what point will the chunk of wood stop moving? Explain in terms of forces acting on the chunk of wood.

(h) Following a line of argument exactly parallel to that used in connection with the chunk of wood in water, describe how helium or hot air balloons rise in air. Be sure to draw the relevant force diagrams.

Now suppose you take a stone or piece of iron and hold it submerged in water just as you did with the chunk of wood in the preceding analysis.

(i) Analyze the forces acting on the stone or iron just as you did the forces acting on the chunk of wood in preceding sections. Draw the corresponding force diagrams and answer the corresponding questions. How do you explain the downward acceleration of the stone immediately after you let it go?

(j) How do you explain the fact that the upward force you exert when you hold the stone submerged in water is perceptibly less than the force you exert when you hold it in air, that is, why does the stone feel "lighter" in water than it does in air?

(k) How do you explain the fact that a boat made of iron floats even though a simple chunk of iron sinks?

(l) "Archimedes's principle" is the name given to the following statement: "An object placed in any fluid is buoyed up by a force equal to the weight of fluid displaced by the object." What is the justification for this statement? Explain it in terms of the sequence of thinking you have done in this question.

3.16 Consider the situation in which you push or pull a lawn or tennis court roller over the ground as shown.

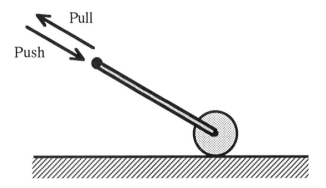

(a) Draw a force diagram for the roller for the case of pushing, a diagram for the case of pulling, and separate force diagrams for the ground in the region of contact with the roller in each case. Describe each force in words and identify the Third Law pairs.

(b) Now examine the following assertion: "It is easier to pull a roller than to push it, but one does a better job of smoothing the ground when one pushes rather than pulls." In the light of the force diagrams you have drawn, do you find this statement to be entirely correct, partly correct, or incorrect? Explain your reasoning.

3.17 A person, exerting a horizontally directed force P, presses a block of mass m against a vertical wall as shown. It is observed that the block does not slide either up or down; it remains at rest in the original position.

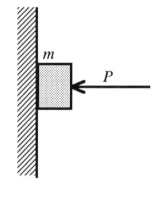

(a) Draw *separate* force diagrams showing (1) the block, with all the forces acting on it, (2) the forces acting on the wall in the region of contact with the block and, (3) the earth and its interaction with the block. Describe each force in words (stating what object exerts this force on what) and identify the Third Law pairs.

(b) How do you account, in your force diagrams, for the fact that the block does not slide down along the wall? Describe three separate changes in this situation each of which would lead, without any other changes, to a condition in which the block starts sliding down the wall.

(c) Suppose m and the coefficient of friction between the wall and the block are known quantities. Describe how you would calculate the value of the force P that is just sufficient to keep the block from sliding. (Your explanation should include an algebraic solution of the problem.)

3.18 A block having a mass of 10.0 kg is pressed against the wall by a hand exerting a force F inclined at an angle θ of 52° to the wall as shown. The coefficient of static friction μ between the block and the wall is 0.20. We shall investigate the question of how large the force F must be to keep the block from sliding along the wall.

There is more physics here than initially meets the eye. Think about the situation in terms of your everyday experience (or better yet, actually try it out): if you start out with a small value of F, the block will tend to slide downward; as you increase F, you reach the point at which the block will no longer slide; as you continue increasing F, the block stays put until, at some larger value of F, it might even begin to slide upward. This is the physics to be investigated, both algebraically and numerically.

(a) First draw well-separated force diagrams of the block and the region of the wall where the two are in contact (1) for the case in which F is small enough that the block tends to slide downward and (2) for the case in which the block tends to slide upward. Denote the various forces by appropriate algebraic symbols; do not put in numbers at this point. (The difference between the two sets of diagrams will reside in the direction of the frictional force.) Describe each force in words and identify the Third Law pairs.

(b) Applying Newton's second law, obtain algebraic expressions for F in terms of mg, μ, and θ for case 1, in which the block is just about to start sliding downward and for case 2, in which it is just about to start sliding upward.

(c) Now put in the various numbers and calculate the value of F for each of the two cases. How large is the spread between the two values? Does your result make physical sense? What is going on at the wall when F lies between the two extremes you have calculated? What happens to the frictional force when F lies between these two extremes?

(d) Return to the algebraic expression for case 2 in which the block is just about to slide upward. What does this expression say happens to F if you keep m and θ constant but increase the value of μ? What is the equation telling us happens at the point at which μ is large enough to make the denominator of the expression equal to zero? Is it possible to make the block slide upward with a sufficiently large F acting at a fixed value of θ regardless of the value of μ? Solve for the value of μ at which it becomes impossible to make the block slide upward, showing that this value depends only on θ and is *independent* of the weight of the block. Do you find this result strange? Why or why not? Could you have anticipated it without having made the mathematical analysis?

3.19 Perhaps you have seen the widely performed demonstration in which a tablecloth is quickly yanked out from under a set of dishes on a table; the dishes remain on the table and are not pulled off to fall on the floor. This is usually described as a "demonstration of the effect of inertia." Let us examine the whole phenomenon carefully.

The figure shows a bowl with mass m_B resting on the tablecloth. A force F is applied to the end of the cloth, and the cloth is yanked out from under the bowl.

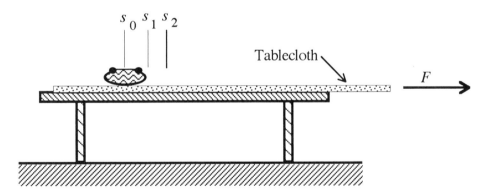

The bowl must, of course, be displaced at least *some* distance along the table. In performing the demonstration, we hope this displacement is small. This is the point we shall examine. Let us subdivide the sliding of the bowl into two obvious stages. In stage 1 the bowl slides from position s_0 (at clock reading t_0) to position s_1 (at clock reading t_1) under the influence of the frictional force exerted by the tablecloth. In stage 2 the bowl slides to a stop at position s_2 (at clock reading t_2) on the table after the cloth is no longer under it. Let us denote the coefficient of sliding friction between the bowl and the cloth by μ_{BC} and the coefficient of sliding friction between the bowl and the table by μ_{BT}.

(a) Draw force diagrams for the bowl in stages 1 and 2, describing each force in words and labeling the various forces with appropriate symbols. Then apply Newton's second law to find the acceleration imparted to the bowl during each stage.

(b) Now that you have expressions for the acceleration during each stage, use the appropriate kinematic equations to obtain expressions for the two successive displacements $s_1 - s_0$ and $s_2 - s_1$ of the bowl. (Do not lose sight of the fact that the bowl is accelerated to a velocity v_1 at instant t_1 and that it coasts to a stop at instant t_2. Note that you can now express $t_2 - t_1$ in terms of $t_1 - t_0$.)

(c) Now show that the total displacement $s_1 - s_0$ of the bowl is given by the following equation:

$$s_2 - s_0 = \frac{\mu_{BC}\, g}{2} \left[1 + \frac{\mu_{BC}}{\mu_{BT}} \right] (t_1 - t_0)^2$$

Note that the time interval appearing in this expression is $t_1 - t_0$, *not* $t_2 - t_1$.

(d) Interpret the equation. What happens to the displacement $s_2 - s_0$ if μ_{BC} is made very large (i.e., the bowl is virtually glued to the cloth)? What happens to the displacement if μ_{BT} is made very small (i.e., the table is virtually frictionless)? Why is the total time interval $t_2 - t_1$ irrelevant? What happens if the cloth to the left of the bowl (i.e., the length that must slide out from under the bowl) is made longer and longer? Thus, all told, under what circumstances does the demonstration ''work'' and under what circumstances does it *not* work?

(e) Now note that the mass m_B of the bowl does not appear in the final expression for displacement and must therefore be irrelevant to the displacement. In what sense can this experiment be a demonstration of the effect of inertia if the inertial mass of the object subject to the effects does not even appear in the final equation for the displacement???

3.20 Consider the two cases shown, in which a magnet holds various objects while you hold the magnet. In diagram (a) the magnet holds a string of iron paper clips hanging end to end; in diagram (b) it holds a small block of iron.

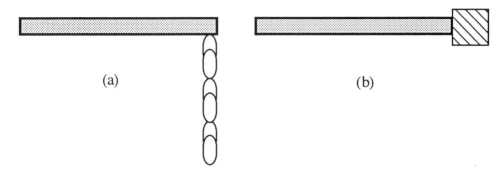

(a)   (b)

(a) Draw separate force diagrams in case (a) for each paper clip and for the magnet. Describe each force in words and identify the Third Law pairs. (Show larger forces with longer arrows and equal forces with arrows of equal length.) How do you account for that fact that the paper clips that are not in direct contact with the magnet do not fall?

(b) Draw separate force diagrams in case (b) for the iron block and the magnet. Describe each force in words and identify the Third Law pairs. How do you account for the fact that the iron block does not fall?

3.21 Suppose we apply the force F_2 to the end of a rope having mass m_R and proceed to accelerate the system, consisting of the massive rope and block A with mass m_A, in the positive x direction along a surface with negligible friction.

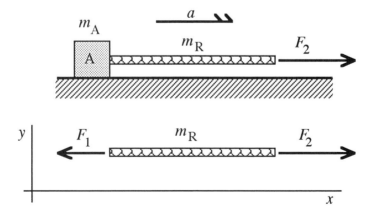

A force diagram for the rope is shown, with the force F_1 drawn smaller than F_2 because the rope is being accelerated and must therefore have a net force acting on it in the horizontal direction. The acceleration of the entire system is denoted by a, since we assume that the rope does not stretch. In the following, we shall concern ourselves with the horizontal forces only and treat vertical forces (e.g., weight of the rope) as unimportant to the physics under consideration.

(a) Draw the force diagram for block A and argue (1) that $F_1 = m_A a$, and (2) that $F_2 - F_1 = m_R a$. How does the magnitude of F_1 compare with the magnitude of F_2? How do you explain the difference physically?

We apply the adjective "tensile" to forces such as F_2 and F_1 acting on a rope, string, rod, bar, or any object being stretched as the rope, in this example, is being stretched. We say that F_1 is the "tension" at the plane cross section through the left-hand end of the rope and that F_2 is the "tension" at the plane cross section through the right-hand end. Similarly, we call the force acting on a plane cross section through *any* location along the rope "the tension at that location." In the light of this definition, we see that tension in the rope is not uniform but varies continuously from one end to the other when the system is being accelerated.

(b) Consider the right-hand half of the rope as a chunk with horizontal forces acting on it. Draw a force diagram of this right-hand half, denoting the force at the left-hand end of this chunk by F_x. Draw a force diagram for the rest of the rope. In the light of our definition, what is an appropriate symbol for the tension in the rope at the left-hand end of the right-hand half? How would you expect this tension to compare in magnitude with F_2 and F_1?

(c) Let us examine the *difference* between F_2 and F_1 as compared with the magnitude of F_2. That is, we shall make use of the results in part (a) to show that

$$\frac{F_2 - F_1}{F_2} = \frac{m_R}{m_R + m_A}.$$

(d) Analyze and interpret this equation: (1) How do the tensions at the two ends of the rope compare (for a fixed acceleration a) if m_R is made smaller and smaller relative to m_A? (2) If m_A becomes indefinitely large relative to m_R? (3) What happens to F_1 as m_A is made smaller and smaller relative to m_R? (4) What is the tension at the left-hand end of the rope if it is being accelerated with the left end free, i.e., with nothing attached to it?

(e) Under what circumstances is tension completely uniform throughout the entire length of a rope? Under what circumstances of m_R compared with m_A may the tension be regarded as *very nearly* uniform while the system is accelerating?

(f) In the light of your answers to the preceding questions, what is the real meaning of the term "massless string" as it is used in many physics problems you have encountered?

Now consider some situations in which a line of discrete objects (that are connected to each other) is being accelerated by a force applied at one end as follows:

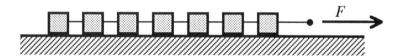

This might be a group of blocks connected by "massless" strings as implied in the diagram. It might be a long chain. It might be a string of freight cars on a railroad track.

(g) Discuss these seemingly different situations qualitatively. In what ways are they basically similar? How do the forces acting on each successive object compare with the forces on the object to the right? How does the force to the right on each successive object compare with the magnitude of the force F ?

(h) Derive an expression for the force to the right on the nth object from the right, assuming the masses of the objects to be identical.

(i) Suppose the line of objects in the preceding figure is connected by identical springs that stretch somewhat in accordance with Hooke's law. If the objects are equally spaced before the system is accelerated (as illustrated), what will the spacing be like after all the objects in the system have acquired the same acceleration?

(j) If, in the situation under consideration in part (i), the force is applied to the lead object abruptly at some instant, will all the objects be accelerated at that same instant? Describe what will actually happen in such a system under these circumstances. You are not being asked to calculate or derive anything; just visualize physically what will happen between the initial instant at which the force is applied and the instant at which all the objects finally have the same acceleration. Under what circumstances will the time interval involved be very short? Under what circumstances might it become quite long?

(k) In the light of your discussion in parts (i) and (j), what do you *visualize* happens in the massive rope after you apply an accelerating force abruptly at one end, even though you cannot actually *see* what happens?

3.22 A heavy rope with total mass m hangs from the ceiling. What is the tension in the rope

(a) at the section at which the rope is fastened to the ceiling?

(b) at the bottom of the rope?

(c) at the middle of the rope?

3.23 Any elastic cord, such as a bungee jumping cord, has an effective spring constant like that of any spring. Will a shorter bungee jumping cord have a larger or a smaller spring constant than a longer cord? Explain your reasoning.

3.24 Suppose we have an object of weight W suspended on a string between poles as shown. We proceed to elevate the object by pulling with forces F on the ends of the string. As we do so, the angle θ decreases, and the string becomes more nearly horizontal.

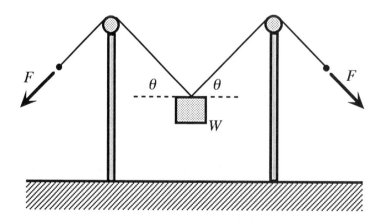

(a) Draw the relevant force diagram, and show that $F = (W/2)\sin\theta$ if we neglect the weight of the string.

(b) Interpret the equation in part (a). What is the value of F when θ is close to 90°? What happens to the magnitude of F as the string becomes more and more nearly horizontal? What is the possibility of getting the string to be absolutely horizontal as long as W is not equal to zero?

(c) In the light of the analysis in part (b), and recognizing that no string can be completely massless, comment on the following little rhyme:

"No force, however great,
  Can stretch a thread, however fine,
  Into a horizontal line
  That shall be *absolutely* straight."

Is this nonsense or is it physically correct? Explain your reasoning.

3.25 Two blocks of dry ice with obviously different masses are sliding at the same velocity on a glass plate and staying a fixed distance apart, as shown in the upper part of the diagram. The motion is very nearly frictionless, since the blocks are sliding on the layer of carbon dioxide gas between their bottom surfaces and the surface of the glass plate. Forces P, of identical magnitude, are suddenly and simultaneously applied to both blocks as shown in the lower part of the figure.

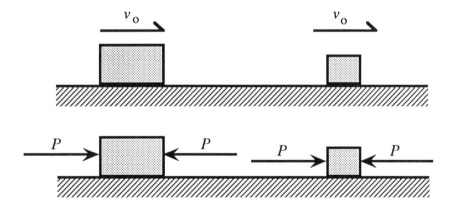

(a) Describe what happens to the motion of each block. Do the blocks continue to stay the same distance apart? If their motions are different after application of the forces, how do they differ? Explain your reasoning.

3.26 A cart with mass m moves to the right on a horizontal surface under conditions of negligible frictional resistance. The velocity of the cart is denoted by v_0 and its position is $x = 0$ at clock reading $t = 0$. At $t = 0$, the force P is suddenly applied at an angle θ as shown. The force remains constant in magnitude and direction after it is applied.

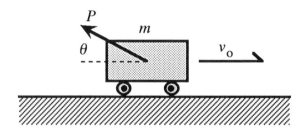

(a) Draw separate force diagrams of the cart and the region of the surface in contact with the cart.

(b) Applying Newton's second law, solve algebraically for the force exerted by the surface on the cart. (Be sure to indicate explicitly your choice of positive direction.)

(c) Applying Newton's second law, solve algebraically for the acceleration imparted to the cart.

Suppose the cart has a mass of 15.0 kg, the initial velocity to the right is 5.8 m/s, the magnitude of the force P is 18.0 N, and the angle θ is 27°.

(d) Using the algebraic result you obtained above, calculate the magnitude of the normal force exerted by the surface on the cart, being careful to give only the proper number of significant figures in the final result. Interpret the result: How does the normal force compare in magnitude with the weight of the cart? Does your result make physical sense? Why or why not?

(e) Using the procedures developed in earlier study of kinematics, calculate where the cart will be located 12 s after the force P is applied. Interpret the result; i.e., describe the motion and successive positions of the cart between clock readings $t = 0$ and $t = 12$ s.

(f) Suppose the cart had a mass of 0.50 kg instead of 15.0 kg. What would happen on application of force P of 18.0 N? Explain your reasoning.

3.27 A force F, acting as shown, imparts an acceleration a to the system consisting of cart A and block B. Block B simply rests against the wall of the cart; it is not fastened to the cart. The coefficient of static friction between the surfaces of A and B is denoted by μ.

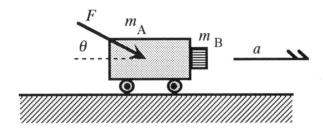

(a) Draw separate force diagrams for bodies A and B and describe each force in words.

(b) The problem to be investigated is that of how large the acceleration a must be to keep block B from sliding downward. Analyze the situation and comment on the feasibility of achieving the condition indicated, assuming realistic values of relevant parameters. (Note: Not all the parameters indicated in the diagram are necessarily relevant to the analysis.)

3.28 A pendulum hangs from the roof of a car that is accelerating to the right as shown. Under these circumstances, the pendulum hangs as shown.

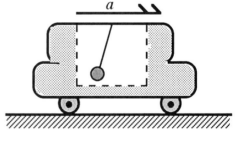

(a) Draw well-separated force diagrams for the bob, the string, and the region of the car roof where the string is attached. Describe each force in words and identify the Third Law pairs.

(b) Show how the pendulum would hang if the car were moving to the right at high constant velocity. Explain your reasoning.

(c) Show how the pendulum would hang if the car were slowing down while moving to the right. Explain your reasoning.

(d) Show how the pendulum would hang if the car were slowing down while moving to the left.

3.29 A car is traveling away from us on a banked road having a curve of radius R. The car is traveling at a speed *greater* than that for which the banking angle θ has the optimum value.

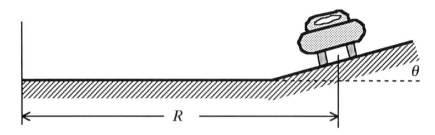

(a) Draw well-separated force diagrams for the car and for the road surface at the location of the car. Describe each force in words and identify the Third Law pairs. (Confine yourself to forces that lie in the plane shown in the diagram; do not try to include forces that act into or out of this plane.)

(b) Draw the same force diagrams as in part (a) for the case in which the speed of the car is *less* than that for which θ is the optimum angle of banking.

(c) Draw the same force diagrams as in part (a) for the case in which the speed of the car is *equal* to that for which θ is the optimum angle of banking.

(d) Describe in your own words the physical differences among the three cases you have illustrated. Pay particular attention to the role of frictional effects parallel to the road surface.

3.30 Consider the situation in which a bob on a string is caused to revolve in a circle of radius r around a fixed point O as shown. The bob has mass m and instantaneous angular velocity ω. The motion takes place in a *vertical* plane.

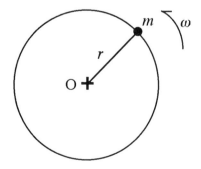

(a) First consider the situation at the very top of the circle. Draw force diagrams for both the bob and the string at that instant. Apply Newton's second law to the motion of the bob and show that if the force exerted by the string on the bob is denoted by T and positive direction is taken in the direction of the centripetal acceleration,

$$T = mr\omega^2 - mg$$

(b) Interpret this relation by examining what it tells us happens to T as we imagine starting with a large value of ω and decrease ω continuously. What is happening when $T = 0$? How does the bob behave when ω is smaller than the value that makes $T = 0$? How might you interpret the negative values of T that are indicated when ω becomes sufficiently small? What will happen if we keep making ω larger and larger without limit?

(c) Examine the situation and behavior of the bob at the bottom of the circle following a sequence parallel to that outlined in parts (a) and (b).

(d) Now, in a parallel analysis, examine what happens to a roller coaster car as it goes over the *top* of a circular track and as it passes through the bottom of a circular track. What force replaces the effect of the string in the preceding analysis? What happens at the top of a circle if the car is going too rapidly?

(e) Apply a parallel analysis to what happens to an object going around a vertical loop-the-loop. How does the situation at the top of the loop-the-loop differ from that at the top of the roller coaster? In what way is the situation at the top of the loop-the-loop similar to that of the bob on the string? In what way is it different?

3.31 A cylindrical chamber in an amusement park rotates around a vertical axis as shown in the following diagram. When the angular velocity is sufficiently high, a person leaning against the wall can take his or her feet off the floor and remain "stuck" to the wall without falling.

(a) For these circumstances, draw force diagrams of the person and for the region of the wall in contact with the person. Describe each force in words and indicate the Third Law pairs.

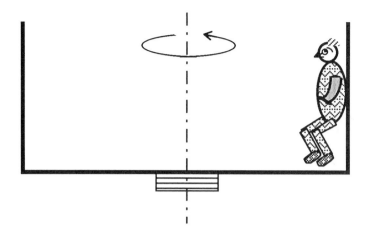

(b) Take the coefficient of friction between the person and the wall to be 1.2. Assume a reasonable mass for the person. Investigate what the period of rotation of the chamber would have to be for the person to be able to take feet off the floor at various chamber radii without falling. Select a combination of values of period and radius that you would consider reasonable and feasible and justify your selection. Explain your reasoning throughout. Would it be easier to produce the desired effect if the coefficient of friction were larger? How large would you like it to be?

(c) Suppose the person were to put a basketball against the wall at the same height as his head while the system is rotating and the feet are off the floor. Draw a force diagram for the ball and describe each force in words. How will the ball behave? Will it stay where it is placed or will something else happen? Explain your reasoning.

3.32 A person, with mass m_p, stands at the rim of a merry-go-round holding a pendulum bob, of mass m_b, on a string. The merry-go-round has a radius R and rotates at a constant angular velocity ω. As the rider adopts the most comfortable stance under the circumstances, a photograph shows him or her to be leaning in such a way that the body line is exactly parallel to the string of the pendulum.

(a) Draw force diagrams for the person, the pendulum, and the region of the merry-go-round in contact with the person. Describe each force in words and identify the Third Law pairs.

(b) In terms of the ideas of force and acceleration we have been studying, explain why the person adopts the body angle observed. In what sense is it "most comfortable"? In the final analysis, are any of the parameters mentioned above irrelevant to the explanation you have given? If so, which ones? Explain your reasoning.

(c) Suppose you wish to calculate the actual angle relative to the vertical adopted by the pendulum string in particular circumstances. Which parameters are needed for the calculation and which, if any, are irrelevant? Explain your reasoning.

3.33 Consider a pendulum bob suspended freely from the ceiling or some other support at each of the following locations at the surface of the earth: the North Pole, the equator, and some intermediate latitude. Let us take the earth to be perfectly spherical (we know this is not actually the case) even though it is rotating. Sketch how the string on which the bob hangs would be oriented relative to a radial line from the center of the spherical earth at each of the three locations (a) if the earth were not rotating and (b) with the earth rotating.

3.34 Be able to explain and interpret the following experiment described by Newton in the *Principia* (words or phrases in brackets [ ] are our editorial insertions to assist the modern reader):

"I tried the thing in gold, silver, lead, glass, sand, common salt, wood, water, and wheat. I provided two equal wooden boxes. I filled one with wood, and I suspended an equal weight of gold (exactly as I could) in the center of oscillation of the other. The boxes, hung by equal threads of 11 feet, made a couple of pendulums perfectly equal in weight and figure, and equally exposed to the resistance of the air. Placing the one by the other, I observed them to [swing] together forwards and backwards for a long while with equal vibrations. And therefore the quantity of matter [inertial mass] in the gold was to the quantity of matter in the wood as the action of the motive force [gravitational force] upon all the gold to the action of the same upon all the wood; that is, the weight of one to the weight of the other."

(a) What was the point of this experiment? What did Newton observe? What did he infer from the results?

3.35 Explain the basis for acceptance of the model according to which, in the solar system, the earth and planets all revolve around the sun rather than the model in which all the other members revolve around the earth. In other words, in what sense do we come to accept the heliocentric solar system rather than the geocentric one? Keep in mind the fact that we do *not* completely reject and abandon the geocentric model; we use it continually, for example, when we navigate on the surface of the earth. [This is not a simple question with a short, pat answer. It involves a lengthy story of successes stemming from a physical *theory*. We do *not* establish the model through direct observation.]

3.36 Suppose that, in a hypothetical other world, the law of gravitation takes the form $F_{grav} = K\sqrt{m_1 m_2}/r^2$ while Newton's laws of motion are otherwise valid, i.e., $F_{net} = ma$. How would falling bodies behave on a planet in this world: Would objects of smaller mass fall with the same acceleration, smaller acceleration, or greater acceleration than bodies of larger mass? Explain your reasoning.

3.37 Consider the case of a satellite of mass m_S in circular earth orbit. Any satellite orbiting in the upper reaches of the atmosphere is subject to a drag force D through collision with molecules of gas that are present in the tenuous upper atmosphere. Under the influence of the drag force, the satellite tends to spiral slowly in toward smaller orbital radii. In the process, the orbital velocity of the satellite *increases*. How can a body speed up under the influence of a retarding (drag) force? What is the *effective* force that must be accelerating the satellite under these circumstances? We shall analyze this situation in detail and arrive at the surprising conclusion that the force accelerating the satellite is actually equal in magnitude to the drag force D! The overall effect is as *though* we turned D around and made it accelerate the object whose motion it is retarding.

The situation we are analyzing is sketched in the diagram on the following page. The diagram also shows separate velocity and force diagrams for the satellite. The mass of the earth is denoted by m_E. The tangential velocity of the satellite at orbital radius r is denoted by v_T. The gravitational force exerted by the earth on the satellite is denoted by F_r.

Let us first note, qualitatively, the effect of the drag force as sketched in the velocity diagram. The drag force D, opposing the motion, causes the satellite to spiral inward. Thus the satellite acquires a very small radial velocity of magnitude v_r, which is greatly exaggerated in the velocity diagram to make the effect visible. With this very small radial velocity, the satellite acquires a total vector velocity of magnitude v and follows a descending path, the tangent to which makes the small angle θ with the direction of the tangential velocity (magnitude v_T) in the circular orbit that would have been followed in the absence of drag. We shall call θ the "angle of descent." The drag force D lies along the line of descent, in the direction opposite to v, as shown in the force diagram. Since v_r is vastly smaller than the tangential velocity v_T, we shall keep using v_T as an adequate approximation to the magnitude of the total velocity v. Thus the application of the drag force causes the satellite to settle down into an angle of descent θ with a radially inward velocity v_r. Since v_r, θ, and the acceleration a_D along the line of descent all depend on D, our problem is to solve for these three relationships.

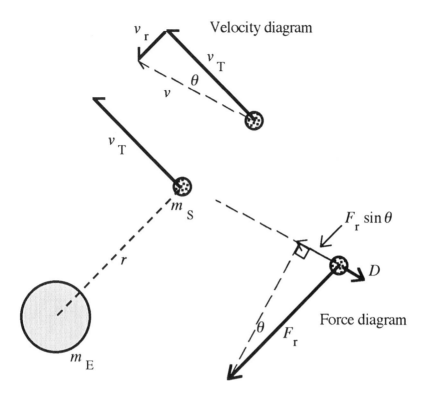

Now let us turn to the force diagram. The total force acting on the satellite is actually the resultant of the drag force D and the centripetal gravitational force F_r. Since D, however, is vastly smaller than F_r, we can see what is happening without trying to construct the resultant but just by looking at the effect of F_r along the line of descent of the satellite. Note that F_r has a nonzero component $F_r \sin \theta$ along the line of descent and in the direction of motion of the satellite. The resultant acceleration a_D imparted to the satellite along the line of descent must be imparted by the force $F_r \sin \theta - D$ acting along this line. In our analysis of what is happening we must study this force.

(a) First show how we arrive at the following two basic equations and describe the physical meaning of the quantity $m_S a_D$ in your own words:

$$F_r \sin \theta - D = m_S a_D \qquad (1)$$

$$F_r = \frac{m_S v_T^2}{r} = G \frac{m_S m_E}{r^2} \qquad (2)$$

We can find out more about a_D and its relation to the various forces and velocities by powerful use of the chain rule for differentiation:

$$a_D = \frac{dv}{dt} = \frac{dv}{dr}\frac{dr}{dt} = v_r \frac{dv}{dr} \cong v_r \frac{dv_T}{dr} \tag{3}$$

where we have made use of the facts that $v_r = dr/dt$ and that v is very nearly equal to v_T.

(b) We can now make use of eq. (2) to show that:

$$m_s a_D = (1/2) F_r \sin\theta \tag{4}$$

Fill in the steps leading to eq. (4) [Hint: Differentiate eq. (2) to find dv_T/dr and use the velocity diagram to show that $v_r \cong -v_T \sin\theta$. (The negative sign is necessary because v_r is negatively directed.)]

(c) By eliminating $F_r \sin\theta$ from eqs. (4) and (1), show that

$$m_s a_D = D \tag{5}$$

and also show that

$$F_r \sin\theta = 2D \tag{6}$$

$$\sin\theta = \frac{2D}{F_r} \tag{7}$$

$$v_r = -2D\sqrt{\frac{r}{m_s F_r}} \tag{8}$$

(d) Interpret the results obtained in part (c) both in your own words and by sketching a force diagram that shows the relationships. Check whether or not the results are dimensionally consistent. What is the point of obtaining eqs. (7) and (8) in addition to (5)? What is the role of the gravitational force (in addition to that of the drag force) in determinining the direction of tangential acceleration? What happens to both v_r and θ as the drag force is increased? Does this make physical sense? Why or why not?

3.38 Suppose two carts, connected by a rope, are being accelerated by a force F as shown. The carts have very different masses, but the frictional forces acting on them are nearly the same. The rope is, of course, under tension in these circumstances.

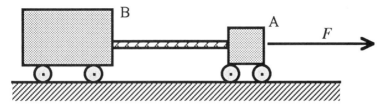

(a) Suppose the force F remains unchanged, but the order of the carts is reversed, i.e., B is placed on the right and A on the left. How does the tension at the center of the rope now compare with what it was before the reversal of the carts? That is, is the tension larger than, less than, or the same as it was initially. Explain your reasoning.

3.39 The law of universal gravitation is given by the equation $F_{grav} = Gm_1 m_2 / r^2$, but the expression for the gravitational force exerted by the earth on an object near its surface (the "weight" of the object) is mg. What has happened to r? Why is r absent from the latter expression?

3.40 Consider the situation illustrated here: the system accelerates when object B is released. The inertial effects of the both the string and the pulley are negligible. The string is essentially unstretchable.

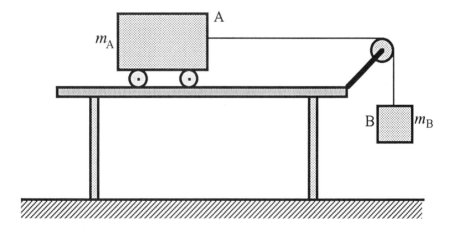

(a) How does the acceleration of body A compare with that of body B? Are the accelerations equal or different? If different, which is larger? Explain your reasoning.

(b) While the system is accelerating, how does the magnitude of the force exerted by the string on cart A compare with the weight of body B? Is the force exerted by the string on cart A equal to, greater than, or smaller than the weight of B? (An algebraic solution is not being called for; the reasoning should be performed qualitatively.) Explain your reasoning.

(c) Suppose that m_B is increased while m_A remains unchanged. What will happen to the acceleration of the system? Will it increase, decrease, or remain unchanged? What will happen to the force the string exerts on body A? What will happen as m_B is increased further and m_A becomes very small relative to m_B? Explain your reasoning.

(d) Describe what will happen to the acceleration of the system and to the force the string exerts on m_A if the changes in part (c) are reversed and m_B becomes very small relative to m_A. Explain your reasoning.

(e) Suppose carts A and B are connected by a rope with significant mass rather than by a massless string. Will the acceleration of the system be uniform or nonuniform? If nonuniform, will the acceleration increase or decrease after release of cart B? Explain your reasoning.

(f) What difficulties would be introduced into the problem if the string or rope connecting the two bodies were stretchable rather than unstretchable?

3.41 Suppose you were to set up a seesaw in an elevator that could be accelerated either up or down.

(a) Imagine first an experiment in which you have a partner of mass equal to your own, and you balance the seesaw while the elevator is at rest. Now, with the seesaw initially balanced, the elevator is accelerated upward. Will the seesaw remain balanced during upward or downward acceleration? Explain your reasoning, being sure to draw separate force diagrams for yourself, your partner, and the seesaw.

(b) Now imagine a second experiment in which your partner's mass is *smaller* than yours. You conduct the same experiment in the accelerating elevator starting with a *balanced* seesaw. Will the seesaw remain balanced during upward or downward acceleration? Explain your reasoning in a manner similar to that in part (a), being sure to draw the relevant diagrams.

3.42 Suppose a cart is accelerating freely down an inclined plane as shown. Inside the cart is a horizontally fixed platform on which rests a block of mass m.

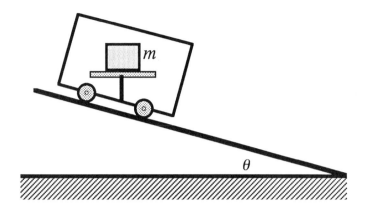

(a) Suppose the platform is frictionless. Will the block appear to stay in a fixed position within the cart or will it slide relative to the platform? If it shifts its position within the cart, which way does it shift? If the block tends to slide on the frictionless platform, in what direction must a frictional force act to keep it from sliding? Explain your reasoning.

(b) Suppose the block is resting on a platform balance while the system is accelerating. (There is sufficient friction between the block and the surface of the platform balance to keep the block from shifting horizontally relative to the cart.) How will the reading on the balance compare with the weight of the block? That is, will it be equal to, greater than, or smaller than the weight of the block? Explain your reasoning.

(c) How will the reading on the balance in part (b) compare with the magnitude of the normal force that would be exerted by the plane on the block if the block were simply sliding down the plane in the absence of friction?

(d) Analyze the situation in part (b) algebraically, and show that the horizontal frictional force must be equal to $mg\sin\theta\,\cos\theta$ and that the reading on the platform balance must be equal to $mg\cos^2\theta$.

3.43 Along with all the other objects at rest at the surface of the earth, you make a complete rotation around the earth's axis in 24 hours. Is it legitimate to say that you and the other objects are "in orbit" around the earth's axis? Why or why not? (In the course of your explanation, be sure to examine what "being in orbit" *means*.)

3.44 Can a satellite, which is to be propelled into earth orbit, be put into an orbit that lies in the same plane as the one in which you rotate around the earth's axis in the position at which you are located on the surface of the earth? Why or why not? Sketch a relevant diagram to assist your explanation.

3.45 Which of the following facts provide supporting evidence for Newton's hypothesis that the force of gravity is proportional to the masses of the interacting bodies?

(a) Kepler's laws are obeyed by planets of very different masses.

(b) The period of an earth satellite is independent of its mass.

(c) Falling bodies all have the same acceleration when air resistance is negligible.

(d) Each of the foregoing facts provides some evidence.

(e) None of the facts (a), (b), or (c) provides evidence for this aspect of the interaction.

3.46 Here object A has four times the mass of object B. The small object C is instantaneously located, relative to A and B at the position shown. Which arrow in the diagram best shows the *direction* in which C would be accelerated by the gravitational forces exerted by A and B at the instant under consideration . Explain your reasoning.

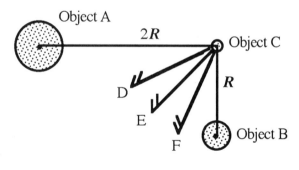

(a) Arrow D.

(b) Arrow E.

(c) Arrow F.

(d) Arrow D, E, or F, depending on the direction and magnitude of the instantaneous velocity of C.

(e) Arrow D, E, or F, depending on whether A and/or B are also free to move.

3.47 Consider each of the following statements. If you believe a statement is consistent with Newton's laws of motion, mark it with Y for "yes" and give a specific example of a situation of the kind described. If you believe a statement is not consistent with Newton's laws, mark it with an N for "no" and explain what is wrong with it.

(a) A body exerts two different forces on another object.

(b) The earth exerts a force on an object in outer space and the object exerts an equal and opposite force on the earth.

(c) A body moves at uniform velocity with only one force acting on it.

(d) A body in outer space accelerates under the influence of a force but exerts no forces on any other object.

(e) During the interval of collision between a small car and a large truck, the small car is subjected to a larger force than is the large truck. This difference accounts for the greater damage sustained by the small car.

(f) Applying the brakes cannot stop a car because the brakes exert a force internal to the car and internal forces cannot accelerate an object.

CHAPTER 4

Momentum and Energy

4.1 A particle of mass m has an initial momentum vector mv_1 as shown. After being given a sharp blow, the particle has a final momentum vector mv_2.

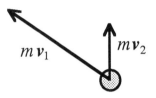

(a) Use the two arrows to construct a vector representing the impulse I that must have been delivered to the particle by the blow that was imparted. Explain your reasoning.

4.2 Two frictionless pucks A and B moving on an air table have velocities v_{1A} and v_{1B}, respectively. The pucks collide and stick together in a perfectly inelastic collision. Puck A has just twice the mass of puck B.

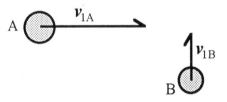

(a) Construct two vectors representing the *momenta* of the two bodies just before collision.

(b) Making use of the vectors in part (a), construct a vector that shows the momentum of the system of the two pucks just *after* the collision. Explain your reasoning.

(c) What *impulse* was imparted to the system (consisting of the *two* pucks) between the initial and final conditions? Explain your reasoning.

4.3 Consider two carts with masses m_A and m_B, respectively, equipped with very soft spring bumpers that undergo quite large displacements as they are compressed in a rectilinear collision. (A similar situation can be set up using gliders on an air track.) When the extra load is removed from cart A, the masses of the two carts are identical. The velocity magnitudes of the carts or gliders before collision are denoted by v_{A1} and v_{B1}, respectively. The purpose of the very soft spring bumpers is to allow one to see the actual compression and springing apart as collision occurs. In performing the following experiments, make the collisions fairly gentle so as not to ruin the bumpers.

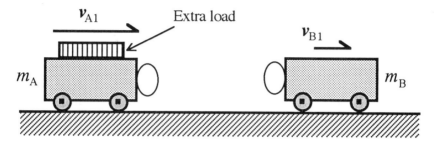

(a) Perform some *qualitative* experiments with a system of this kind, starting with body B stationary ($v_{B1} = 0$), and observing what happens under various circumstances without making numerical measurements: (1) What happens when neither cart is loaded and $m_A = m_B$? (2) What happens when cart A is loaded so that $m_A > m_B$? (3) What happens when cart B is loaded and $m_A < m_B$?

(b) What happens when the carts approach each other with approximately equal but opposite velocities?

(c) In at least a few of the experiments, sketch what the graph of force versus clock reading must look like for each cart over the interval from just before to just after contact. Directly underneath this graph and using the same scale for clock readings, sketch what the acceleration and velocity graphs must look like. Then sketch what the corresponding graphs for collisions between billiard balls or steel ball bearings (where it is impossible to see the deformations that take place) might look like on a similar time scale. What determines the time interval of interaction in these widely different circumstances?

(d) Now consider the collision between carts or gliders equipped with strong magnets mounted so as to repel each other. Carry out observations of collisions, such as those in part (a), if at all possible. Make the collisions quite gentle so that the magnets do not actually strike each other but the carts still spring apart. (Even though the carts never touch each other, we still call this a collision.) What is happening in the absence of physical contact? What do the various graphs look like in such cases? What is the time interval of interaction, and how is this represented on the graphs?

(e) Suppose the magnets were mounted so as to attract rather than repel each other (or imagine the colliding objects to be carrying unlike electrical charges). Describe what might happen in some such collisions and sketch corresponding graphs of force versus clock reading. How might you rig a purely mechanical experiment in which the colliding objects stick together instead of springing apart?

(f) How would a significant amount of friction influence some of the cases you have been observing?

Note to the student: Question 4.4 illustrates an approach to the idea of conservation of momentum utilizing transformation of frames of reference. This approach provides valuable practice preparatory to study of the theory of relativity. In the latter study you will have occasion to make such transformations repeatedly.

In elastic rectilinear collisions of bodies with equal masses, it is readily observed that (1) when the bodies approach each other with equal and opposite velocities, they rebound with equal and opposite velocities (i.e., they exchange their initial velocities); and (2) when the first body is moving toward the right while the second body is stationary, the first body stops still and the second body moves off with the same velocity as the first (i.e., the bodies again exchange their initial velocities).

Christian Huygens, Newton's great Dutch contemporary in the seventeenth century, raised a penetrating question about these two observations: Are these simply two unconnected and independent phenomena, or is there a deeper order in nature that connects them to some common law or principle? Let us follow his simple and powerful analysis of the problem.

4.4 We start, as Huygens did, with the observation that ''equal hard bodies'' (meaning perfectly elastic bodies of equal mass) approaching each other with equal and opposite velocities rebound with velocities unchanged in magnitude as illustrated in the following figure. Our frame of reference (set of coordinate axes) is denoted by O. The two bodies, A and B, moving parallel to the x-axis with velocities v_{A1} and v_{B1}, respectively, undergo a rectilinear collision.

Not only is this starting point consistent with observations but it is also consistent with our deep sense of symmetry in natural phenomena. It is this same sense of symmetry, for example, that leads us to expect identical objects, placed at equal distances from the pivot point of a seesaw, to balance each other.

We would be surprised at any other outcome and would be very certain that the two objects were not identical if they failed to balance.

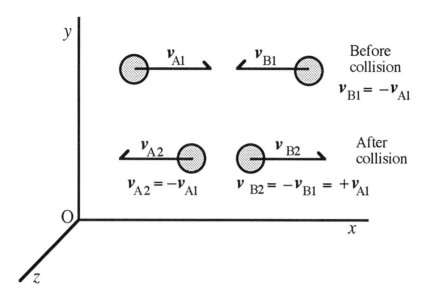

We next ask the question: What would we expect to happen if body B is initially stationary, and an identical body A approaches with velocity v_{A1} in the same reference frame O as shown in the following figure.

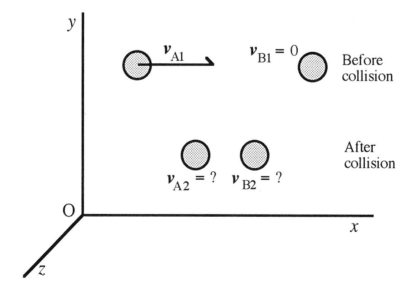

Huygens proceeded to address this question through the clever device of viewing the second collision from another frame of reference O', which made the second collision identical in character with the first one. He chose O′ to move to the right at uniform velocity $v_o = v_{A1}/2$ relative to frame O as follows.

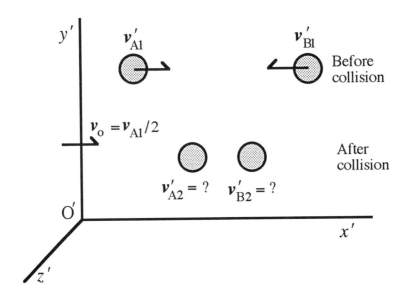

(a) Argue in your own words that if we take the velocity v_o of frame O′ relative to O to be $v_{A1}/2$, the two objects, viewed from O′, will appear to be approaching each other at equal and opposite velocities of magnitude $v_{A1}/2$.

(b) Still from the point of view of O′: the two bodies will now rebound with equal and opposite velocities, also of magnitude $v_{A1}/2$. Enter these after-collision velocities in O′ on the figure, with appropriate symbols.

(c) Now view these after-collision velocities from the original frame O. Show that from the point of view of O, body A will have zero velocity while body B will be moving to the right with velocity v_{A1}.

(d) Argue in your own words that this analysis shows the two seemingly different types of collision to be intimately related and to be governed by some common underlying order in nature. (This common order turns out to be conservation of momentum, regardless of frame of reference.)

4.5 Carry out an analysis of the perfectly *inelastic* collision of two identical bodies by following a sequence exactly similar to the one used in the preceding problem: take as initially given the perfectly symmetrical case in which the two bodies, approaching each other with equal and opposite velocities, come to a dead stop on collision. Then consider the problem in which body A moves to the right at velocity v_{A1} in frame O and collides with stationary body B; the two bodies stick together and move toward the right at a final velocity v_2.

(a) Now view this latter collision from a frame of reference O′ in which the two bodies appear to approach each other at equal and opposite velocities and come to dead stop. (What must be the velocity of O′ relative to O?) Sketch the frame and label the velocities as they appear in it.

(b) Now view the final situation in O′ from the original frame O. Show that the velocity of the combination of A and B, stationary in O′, must have the velocity $v_{A1}/2$ in O. Note that this constitutes a prediction of what is actually observed to happen.

4.6 Explain, as though you were making a presentation to your fellow students, the connection between Newton's third law and the law of conservation of linear momentum. Be sure to include the following: an explanation of why it is necessary to define clearly the *system* under consideration; an explanation of the distinction between an open and a closed system; an explanation of the problem that would arise if the forces the interacting objects exert on each other were *not* equal and opposite to each other *instant by instant* throughout the interaction as demanded by Newton's third law. Give an example of some common events in which, at least for a very short time interval, the forces of interaction between well-separated bodies are, in fact, *not* equal and opposite instant by instant.

4.7 Two objects with masses m_A and m_B, respectively, form a closed system and undergo a perfectly elastic rectilinear collision. Object B is initially stationary, and object A has a nonzero initial velocity. Starting with the relevant equations derived in study of such collisions and explaining your reasoning, predict the motion of each body after the collision if

(a) $m_A > m_B$

(b) $m_A = m_B$

(c) $m_A < m_B$

(d) Now examine and predict what happens in the limit in which m_B is vanishingly small relative to m_A.

(e) Do the same for the limit in which m_A is vanishingly small relative to m_B.

(f) Explain the connection between your prediction in part (e) and what happens when a ball bounces elastically from a rigid wall.

4.8 Suppose that in our laboratory frame of reference O_L, a flat, relatively massive steel wall moves to the right (in the x-direction) with uniform velocity of magnitude v_W. The wall is perpendicular to the x-axis. A small steel ball moves toward the right with a larger velocity v_B, catches up with the wall, and undergoes a perfectly elastic collision, bouncing back toward the left.

(a) What is the initial velocity of the ball in the frame of reference O_P of the steel plate? With what velocity will the ball rebound in this frame of reference? What will be the final velocity of the ball in frame O_L? Explain your reasoning.

(b) Suppose now that the wall moves toward the *left* with uniform velocity v_W as the ball still moves toward the right. What will be the final velocity of the ball in frame O_L after the rebound? Explain your reasoning following a sequence similar to that in part (a).

(c) In each of the two preceding cases, what happens to the kinetic energy of the ball in frame O_L: Does it increase, decrease, or remain unchanged? Explain your reasoning. If the kinetic energy of the ball changes, where does any decrease go and where does any increase come from?

(d) A piston, confining a gas in a cylinder, is moving, at uniform velocity, either inward (compressing the gas) or outward (expanding the gas). In the light of your analysis in part (c), describe what must be happening, on the average, to the kinetic energies of molecules of gas that rebound from the piston as the gas is compressed and as it is expanded .

(e) In the situations visualized in part (d), take the system under consideration to be the gas. Explain where the predicted kinetic energy changes go or come from. (What happens in the way of work being done by the piston on the gas or by the gas on the piston?)

4.9 Suppose we take a toy balloon, blow it up with air, and let it go without tying up its mouth. It is a familiar experience that as the balloon deflates, it flies erratically around the room until it is completely deflated. (In the following, be sure to define the system you will discuss and indicate whether it is open or closed.)

    (a) Without equations or formulas, but using the concepts of impulse and change of momentum, describe, *qualitatively*, the behavior of the balloon while it is deflating.

    (b) Describe the energy changes that take place during the same interval, starting with the situation in the inflated balloon.

4.10 If one wants to jump from a height without getting hurt, one chooses to jump into a stretchable net or pile of hay or pile of soft mattresses rather than hitting the hard ground.

    (a) Explain this choice in terms of the impulse–momentum theorem and the concept of "average force." How does it come about that the risk of injury is very much less in the softer cases than on the hard ground, while the momentum change is exactly the same in all instances?

    (b) Describe the energy changes that take place in the various circumstances.

4.11 Two objects, one with a large mass m_A and the other with smaller mass m_B, are released from rest relative to an observer in free space. The objects then accelerate toward each other under the influence of their mutual gravitational attraction (no other forces are acting). Consider the instant just before the objects collide, when they have the final velocities of magnitude v_A and v_B, respectively. (Explain your reasoning in answering each of the following questions.)

    (a) How does the net impulse delivered to A compare with the net impulse delivered to B?

    (b) How does the momentum change of A compare with the momentum change of B?

    (c) How does the final (just before collision) velocity of A compare with the final velocity of B? (Derive an expression for v_A in terms of v_B and the ratio of the two masses.) How do the displacements of the two bodies compare with each other for various ratios of the masses?

(d) How does the final (just before collision) kinetic energy of A compare with the final kinetic energy of B? Are the two energies equal or is one greater than the other? (Derive an expression for the ratio of the two kinetic energies.)

(e) Was the work done on A equal to the work done on B? How do you explain the fact that the work done on each body is different while the impulse delivered to each body has the same magnitude?

(f) Where did the total final kinetic energy of the system of the two bodies come from?

In his *Astronomia Nova* of 1609, seventy-eight years before Newton's *Principia*, Johannes Kepler made what has become a famous and often quoted remark:

> "If two stones were placed in any part of the world, near each other yet beyond the sphere of influence of a third related body, the two stones, like two magnetic bodies, would come together at some intermediate place, each approaching the other through a distance in proportion to the mass of the *other*." [In other words, the displacements of the two stones would be *inversely* proportional to their masses.]

(g) How does Kepler's prediction compare with your result in part (c)?

Kepler's word for "mass" was then the Latin word "*moles*" — a term denoting bulk of matter in some vague sense rather than our modern, operationally defined concept. Clearly defined conceptions of inertial mass and gravitational mass were then still far in the future (even beyond Newton's *Principia*). Yet Kepler must be given credit for profound, partial insight.

(h) Comment on Kepler's statement in the light of the analysis you have carried out in the first part of this problem: In modern terms, what are its dynamical implications? Is Kepler's prediction consistent with our present knowledge and concepts? Why or why not? What name do we give to the "intermediate place" to which Kepler refers? What connection, if any, do you see between this situation and the one in which two carts on a level table (or two gliders on an air track) have a compressed spring between them and fly apart when the spring is released?

4.12 A glider of mass m moves in the positive direction on an air track (negligible friction) with velocity magnitude v_0. At the end of the track, it makes a perfectly elastic collision with a spring bumper and rebounds with the same magnitude of velocity.

(a) Explaining your reasoning and drawing an appropriate momentum vector diagram, write an algebraic expression for the change of momentum of the glider.

(b) What net impulse must have been delivered to the glider by the spring bumper? (Do not lose sight of the fact that impulse and momentum are both vector quantities.) Explain your reasoning.

(c) What was the kinetic energy change (algebraic expression) of the glider? Explain your reasoning.

(d) What was the net work done by the spring on the glider? Explain your reasoning.

(e) Describe the sequence of energy changes that takes place in the glider–spring system between initial contact of the glider with the spring and the final parting of contact.

(f) Explain how the net work done by the spring on the glider can be zero while the net impulse delivered by the spring to the glider is not zero.

(g) Suppose now that the collision is partly inelastic and the glider rebounds with a speed of 0.9 v_0. Answer questions (a) – (e) for this case.

4.13 A spring is compressed between two carts and is temporarily clamped so that carts and spring move as a single unit. Initially the carts move toward the right on a level surface with a velocity of 2.80 m/s. The masses of the carts are 1.60 kg and 2.50 kg, respectively, as shown. The mass of the spring is very much smaller than that of either cart and can be neglected. At a certain instant the clamp is suddenly released, and the carts separate, with cart B moving to the right at 3.20 m/s. (In each part of the following analysis, be sure to define the system you are dealing with and to indicate what conservation law or laws you are applying.)

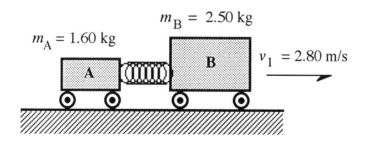

$m_B = 2.50$ kg

$m_A = 1.60$ kg

$v_1 = 2.80$ m/s

(a) Indicate your choice of positive direction, and, explaining your reasoning in neat and intelligible sequence, calculate the velocity of cart A immediately after decompression of the spring.

(b) As a result of this interaction, do you expect the final total kinetic energy of the system (the two carts) to be equal to, greater than, or less than the total initial kinetic energy? Explain your reasoning carefully in terms of the different kinds of energy available in the system. Then check your answer by making the relevant numerical calculations.

(c) Is this event to be described as involving an elastic or an inelastic interaction between the two carts? Explain your reasoning.

4.14 Suppose you throw a stone having a mass of 0.50 kg vertically upward. Let us assume that your hand exerts an average force of 110 N over an arm displacement (upward) of 0.60 m. (Through the following sequence of questions, we shall explore, in terms of the energy and momentum concepts, what happens to the stone, and we shall ascertain whether the numerical values given above are physically reasonable. Follow the sequence carefully for the exercise it provides in using and interpreting the energy and momentum concepts, setting up numerical expressions and indicating your line of reasoning. Do *not* resort to calculating accelerations or using the *kinematic* relations except to check your results for internal consistency.)

(a) Draw and label the following force diagrams: for the stone during the act of throwing, for the stone after it has left contact with your hand, for your own body, and for the ground in the vicinity of your feet. Describe each force in words and identify the Third Law pairs.

(b) Calculate the work done on the stone *by your hand* in the act of throwing.

(c) Calculate the *net* work done on the stone during the act of throwing. Explain why this number differs from the one obtained in part (b).

(d) Calculate the kinetic energy change imparted to the stone in the act of throwing, i.e., the kinetic energy of the stone at the instant it leaves your hand.

(e) Calculate how high the stone will rise (making use of the kinetic and potential energy concepts).

(f) Using the result obtained in part (d), calculate the velocity of the stone at the instant it parts contact with your hand.

(g) Calculate the change of momentum that was imparted to the stone in the act of throwing.

(h) What magnitude of *net* impulse, in what direction, must have been imparted to the stone in the act of throwing? What total magnitude of impulse was imparted to the stone by your hand? Why is the magnitude of this impulse different from that of the *net* impulse?

(i) During the act of throwing, were any net impulse and change of momentum imparted to the system consisting of your body and the earth? If not, why not? If so, what was the magnitude and direction?

(j) Are the numbers given in the problem and the resulting values that you have calculated physically reasonable? Justify your answer on the basis of your own experience in throwing objects upward.

(k) Where does the kinetic energy imparted to the stone come from?

(l) How much work is done by the normal force exerted by the ground on your body?

(m) What is the *rate of change* of its momentum while the stone is on the way up? At the instant it is at the top of its flight? On the way down? Explain your reasoning.

4.15 Suppose two pendulum bobs with known masses m_A and m_B, respectively, are suspended from a rigid support. As implied in the diagram, $m_A < m_B$. Bob A is displaced to position 1 and is let go from rest at that position. The collision between A and B is taken to be perfectly elastic.

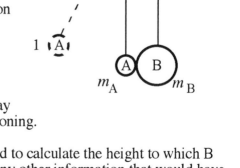

(a) Describe the collision qualitatively: Which way does B move after the impact? Which way does A move? Explain your reasoning.

(b) Describe how you would proceed to calculate the height to which B rises after the collision: Is there any other information that would have to be given or that you would have to assume? What relations would you use and how would you use them? Is it possible to solve the problem in one step with one single equation or is more than one step necessary? Explain your reasoning in detail in your own words as though you were leading a fellow student through the reasoning and the algebraic steps. Be sure to make clear how the concept of "perfectly elastic collision" is being utilized in the reasoning.

(c) Is it possible to calculate how high A rises after the impact? If so, how?

(d) What would you have to know to calculate the initial work put into the system to get it started?

4.16 A small cart starts from rest at point A and rolls down a circular track of radius R. At point B (the lowest point on the track), it enters another circular track, which has a radius $R/2$. Positions of the cart along this latter track can be described in terms of the angle θ measured from the vertical line through B. In the following analysis, treat the cart as a particle moving along arcs of the radii indicated and the motion as frictionless.

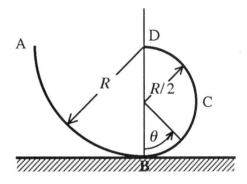

(a) Think carefully about how the normal force N exerted by the track on the cart beyond point B must be varying as the cart ascends the track and argue *qualitatively* (without mathematical analysis) that the cart cannot drop off the track before it reaches point C and that it *must* drop off the track before it reaches point D.

(b) Now proceed to confirm the qualitative argument in part (a) with a mathematical analysis: using a conservation of energy argument, show that the square of the velocity v_B of the cart at point B is given by

$$v_B^2 = 2gR \tag{1}$$

(Be sure to define the *system* to which you are applying the energy argument.)

(c) Using a similar conservation of energy argument, show that beyond point B, the square of the velocity of the cart depends on θ in the following way:

$$v^2 = gR(1 + \cos\theta) \tag{2}$$

(d) Draw a force diagram for the cart at an arbitrary θ and apply Newton's second law to the motion to solve for the normal force N exerted by the track on the cart as a function of θ. Interpret the equation for N, showing that the result confirms the simple argument made in part (a), and find the numerical value of the angle θ at which the cart will drop off the track. [Hint: what value of N will signal the parting?] (Answer: $\theta = 120°$.) Point out the terms that would be affected by the presence of friction and show that the effect would be to decrease the angle at which the cart drops off, as one would expect intuitively.

(e) How does it come about that even in the frictionless situation, the cart cannot get back to its original height R above point B? Is this a violation of conservation of energy? Why or why not?

(f) Identify some kinematic and dynamic variables that change abruptly at point B. Identify some variables that do *not* change abruptly at point B.

4.17 Lord Kelvin describes encountering Joule and his bride honeymooning in Switzerland. Joule was stopping by roadside waterfalls and measuring temperatures of the water in the stream entering the falls and of the water in the still pool at the bottom of the falls.

(a) Describe in good English the energy transformations that occur between the initial and final conditions in which Joule was interested. From what you know about Joule's work and interests, why do think he was bothering with these observations?

(b) Estimate the temperature difference that might arise ideally under these circumstances in the case of a waterfall 50 m high. Explain your reasoning carefully.

(c) Is your estimate of temperature likely to be an upper limit, a lower limit, or something in between? Explain. List the effects that would make the actual temperature difference deviate from the ideal result. What would be the direction of the deviation? Explain your reasoning.

4.18 A satellite moves around the earth in a sufficiently large circular orbit that the frictional resistance to the motion is essentially zero. Explain how we know that under these circumstances, the motion will persist without our putting any additional energy into the system.

4.19 Consider a planet orbiting the sun in an elliptical orbit. Sketch such an ellipse, label some key positions, and then describe *qualitatively* how the kinetic energy of the planet and the potential energy of the planet–sun system vary as the planet executes its orbit.

4.20 Let us analyze and compare the energy changes that take place in systems 1 and 2: the first consists of a frictionless puck of mass m on an air table. The puck is attached to a string, which can be pulled down through a smooth hole in the air table at point O. The puck is set into circular motion at angular velocity ω_0 about point O at an initial radius r_0.

System 2 consists of a planet with mass m_P in circular orbit at angular velocity ω_0 and at radius r_0 around a very massive sun with mass m_S. (With m_S very much larger than m_P, any motion of the sun will be completely negligible, and we need concern ourselves only with the motion of the planet.)

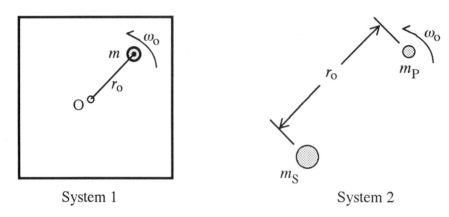

System 1   System 2

Suppose the string in system 1 is pulled slowly down through the hole so that the frictionless puck is drawn in from the initial radius r_0 to a final smaller radius r and corresponding angular velocity ω without acquiring appreciable kinetic energy in *radial* motion. This is like raising an object slowly without imparting appreciable kinetic energy. In raising the object, we calculate the work we do on the object–earth system by taking the force we exert to be essentially equal in magnitude to the weight of the object. In the case of the frictionless puck, the force we exert in pulling the string is essentially equal to the centripetal force instantaneously exerted on the object. (The situation is analogous to the one in which the spinning skater draws in his or her arms and experiences increasing angular velocity while angular momentum is being conserved.)

(a) Treating the puck as a point particle, obtain an algebraic expression for the work W done on the puck in displacing it from the initial radius r_0 to a final radius r subject to the auxiliary condition that the angular momentum $mr^2\omega$ of the puck is conserved. You will find that the final expression can be written in either of two ways:

$$W = \frac{1}{2} m r_0^2 \omega_0^2 \left[\frac{r_0^2}{r^2} - 1 \right] \tag{1}$$

$$W = \frac{1}{2} m r_0^2 \omega_0 \left[\omega - \omega_0 \right] \tag{2}$$

(b) Given the initial and final conditions for r and ω as defined above and the fact that the instantaneous tangential velocity $v_t = r\omega$, argue, independently of the relations (1) and (2) for W, that the change in kinetic energy (ΔKE) of the puck must be given by

$$\Delta\text{KE} = \frac{1}{2} m r^2 \omega^2 - \frac{1}{2} m r_0^2 \omega_0^2 \tag{3}$$

(c) Now show that either one of the expressions (1) or (2) for W is in fact equal to expression (3) for ΔKE. How do you interpret this equality? That is, what happened to the work done in pulling the string and decreasing the radius from r_0 to r? Does this make sense in terms of energy conservation? Why or why not?

(d) Now suppose that in system 2, the planet is slowly lowered in toward the sun just as the puck was pulled in toward the center of its circle. (This is like slowly lowering an object in the earth's gravitational field.) Show that the work W that must be done *on* the planet–sun system to effect this displacement is given by

$$W = -G m_p m_s \left[\frac{1}{r} - \frac{1}{r_0} \right] \tag{4}$$

Explain the meaning of the minus sign and argue that this expression is, in fact, the decrease in potential energy of the planet–sun system for the specified change in radial position of the planet.

(e) Treating the planet as a point particle, argue that the change ΔKE in its kinetic energy must be given by an expression identical to (3) for the frictionless puck, simply replacing m by m_p. Then show that this expression can be reduced to the form

$$\Delta\text{KE} = \frac{1}{2} G m_p m_s \left[\frac{1}{r} - \frac{1}{r_0} \right] \tag{5}$$

(f) Note that, in the case of system 2, the work done on the planet–sun system in changing the radial position of the planet is *not* equal to the change in the kinetic energy of the planet, while in system 1 the work done in changing the radial position of the puck *is* equal to the change in the puck's kinetic energy. How do you account for this profound difference between the two systems? Is this a violation of the law of conservation of energy ? Why or why not? (In the gravitational case, what must happen to the difference between the decrease in potential energy of the system and the increase in kinetic energy of the planet?) Under what circumstances, in the gravitational case, might the radial position of a planet or satellite be caused to increase or decrease? (In your comparison of systems 1 and 2, note that in system 1, the string exerts a passive force that adjusts itself to the given conditions and that any angular velocity is possible at any radius as long as the string doesn't break. However, in system 2, the law of gravitation requires that a given value of angular velocity is possible at only one particular radius.)

(g) Compare system 2 with the situation in which a small negatively charged particle with mass m_P and charge $-q_P$ revolves in a circular orbit around a massive positively charged particle with mass m_S and charge $+q_S$. Discuss what effects might lead to an increase or decrease in the radial separation in this system.

(h) Return to Question 3.37 if you worked on it previously. How do you account for the fact that the satellite, under the influence of a drag force, gains, rather than loses, kinetic energy?

4.21 Suppose that two carts, with a compressed spring between them, are free to roll on a level table. The carts have very different masses and are initially stationary; consider friction to be negligible. The spring is released, and the carts fly apart. How do the kinetic energies of the carts compare at the instant they part contact with the spring? That is, are the kinetic energies equal or is one larger than the other? To answer this question, choose your own symbols and analyze the situation algebraically, explaining your reasoning and interpreting the final result.

4.22 A coil spring obeying Hooke's law has a spring constant denoted by k. Suppose, by applying a force denoted by F, we stretch the spring by a length ΔL from its relaxed position .

(a) Sketch a graph of applied force versus spring extension from 0 to ΔL.

(b) Appealing to the definition of "work," explain why areas under this graph, for different extensions ΔL, represent amounts of work done on the spring by the applied force. Explain what happens to the work that is done on the spring.

(c) Shade on the graph an area that represents the amount of work done in stretching the spring from extension 0 to $\Delta L/4$. Use a different shading to indicate the area that represents the work done in stretching the spring from extension $\Delta L/2$ to $3\Delta L/4$. Explain why the amount of work done in the second case is different from that done in the first, even though the change of length of the spring is the same in both cases.

4.23 A pendulum bob on a string of length L is elevated from its lowest position to the point at which the string makes and angle of 90° with the vertical. Take the potential energy of the bob–earth system to be zero when the bob is at this level. The bob is now released, swinging from the 90° to the 0° position.

(a) Compare the magnitude of change in potential energy of the system between the 90° position and the 45° position with the change between the 45° position and the 0° position. Is one change larger than the other or are they equal in magnitude? Explain your reasoning.

(b) How much work is done by the string on the bob during the descent? Explain your reasoning.

4.24 A ball is thrown from ground level with initial horizontal velocity component v_{ox} and initial vertical velocity component v_{oy} and returns to ground level. Neglecting friction and explaining your reasoning in each instance, write expressions in terms of these two velocities for:

(a) The largest kinetic energy of the ball during its flight.

(b) The smallest kinetic energy of the ball during its flight.

(c) The maximum potential energy of the ball–earth system during the flight.

4.25 The following sine function graph shows one complete cycle of the displacement–clock reading history of a simple harmonic oscillation of a block on a spring.

a) On a similar set of coordinates directly below this graph, sketch the corresponding history of kinetic energy versus time for the block. (This is a purely *qualitative* question. No numbers are available; just sketch the shape of the required graph.) Give a brief description of the reasoning you used in making the sketch.

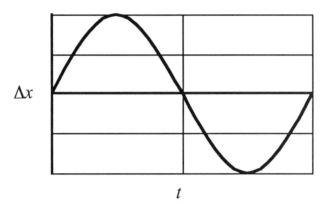

t

Note to the instructor: Questions 4.26 through 4.28 are designed to help students master the meaning and interpretation of algebraic signs that arise in describing the energy changes that take place in several simple cases. Such mastery of algebraic signs can be attained *only* through practice of this kind, and the practice is rarely available.

4.26 Let us consider an object (say a ball) that moves up or down freely in the vertical direction. Air resistance is assumed to be negligible; only the gravitational force is acting; and we take positive direction upward as shown. We treat the symbol g for the acceleration due to gravity as a magnitude only and not as a vector quantity with directional algebraic signs.

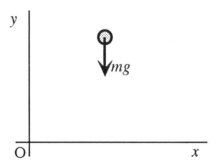

(a) Show that the work–kinetic energy theorem applied to a vertical displacement Δy in this situation takes the form

$$-mg\Delta y = (1/2)mv_2^2 - (1/2)mv_1^2$$

(b) Now consider what happens when you let the ball go from rest at some initial high elevation y_1. What is the value of v_1? Which is the only direction in which the ball will move? What will therefore be the algebraic sign of any possible Δy? What will be the algebraic sign on the left-hand side of the equation? What does the equation say must happen to the kinetic energy of the ball under these circumstances? Does this make sense? Why or why not?

(c) Suppose, instead of letting the ball go from rest, we give it an initial *upward* velocity v_1 and therefore an initial kinetic energy $(1/2)mv_1^2$. What will be the sign of Δy and the algebraic sign on the left-hand side of the equation? Examine what the equation says must now happen to the kinetic energy $(1/2)mv_1^2$ as the ball rises. What does the equation say about how large Δy can become? (Argue that the result loses physical meaning after a certain value of Δy.) What happens after the largest Δy is attained?

(d) Now suppose that the x-axis represents a floor with which the ball can undergo a perfectly elastic collision, reversing its velocity on impact with no loss in magnitude. Show that the equation above says that if the ball is released from rest at some initial height above the floor, it will continue bouncing up and down, always returning to the height at which it was released.

(e) Suppose that the collision with the floor is not perfectly elastic and that the magnitude of the velocity decreases somewhat on each bounce. Show what the equation says will happen under these circumstances.

4.27 Let us consider the case in which an object (say a ball) moves up or down in the vertical direction under the combined influence of its weight mg and a constant upward force of magnitude P as shown. P might be equal to mg, or larger or smaller. The imposed vertical displacement is denoted by Δy.

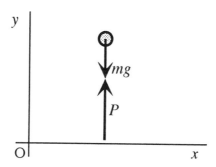

Air resistance is assumed negligible, and we take positive direction upward. We treat the symbol g for the acceleration due to gravity as a magnitude only, not as a vector quantity with directional algebraic signs. In the following analysis, we shall take P, mg, and Δy as specified quantities and ask what happens to the kinetic energy of the ball under various circumstances.

(a) Show that the work–kinetic energy theorem applied to this situation takes the form

$$(P - mg)\Delta y = (1/2)mv_2^2 - (1/2)mv_1^2$$

(b) Suppose first that P is *larger* than mg and Δy is positive. Which way has the ball been moving during the displacement being considered? What is the algebraic sign of the left-hand side of the equation? What must be the final kinetic energy of the ball (in terms of the known quantities) if it started from rest at position y_1? What must be its final kinetic energy if it had velocity v_1 at position y_1? Show that the final algebraic results correspond to what you would expect to happen physically in these circumstances.

(c) Suppose that P is *smaller* than mg and Δy is negative. Which way has the ball been moving during the displacement being considered? What is the algebraic sign of the left-hand side of the equation? What must be the final kinetic energy of the ball (in terms of the known quantities) if it started from rest at position y_1? What must be its final kinetic energy if it had velocity v_1 at position y_1? Show that the final algebraic results correspond to what you would expect to happen physically in these circumstances.

(d) Suppose that P is *smaller* than mg but that Δy is positive. Could the ball have started from rest at position y_1? Why or why not? Which way has the ball been moving during the displacement being considered? What is the algebraic sign of the left-hand side of the equation? What must be the final kinetic energy of the ball (in terms of the known quantities) if it had an upward velocity v_1 at position y_1? Show that your final algebraic result corresponds to what you would expect to happen physically in these circumstances.

(e) Suppose that P is *larger* than mg but Δy is negative. Examine and interpret this situation in the manner that has just been outlined in part (d).

(f) Discuss the situation in which P is just infinitesmally greater or smaller than mg so that the ball *is* displaced up or down but with negligible acceleration.

4.28 A frictionless puck of mass m rests on a level air table. The puck is connected to one end of a spring that has a spring constant k and a mass that is negligible compared to the mass of the puck. The other end of the spring is fastened to the wall as shown in the diagram. The origin of the x coordinates is located at the relaxed position of the left-most end of the spring. In the following analysis and visualizations, let us confine the puck to displacements that do not damage or exceed the capacity of the spring. The free body force diagram of the puck has omitted the balanced vertical forces so as to avoid cluttering the figure.

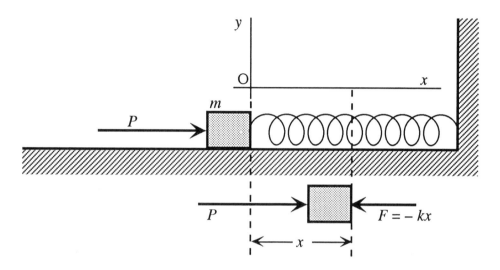

A horizontal force P, which may vary with position, is exerted on the puck, displacing it to the right. The spring obeys Hooke's law, and the opposing force, kx, exerted by the spring on the puck increases linearly with displacement from the relaxed position.

(a) Show that if the force P is applied when the puck is initially at rest at the origin, the work–kinetic energy theorem requires that

$$\int_0^x P(x)\,dx - (1/2)kx^2 = (1/2)mv^2 \qquad (1)$$

where x denotes the final position and v the velocity at that position.

(b) Interpret what eq. (1) is telling us. For example, if the left-hand side of the equation is positive, what must have been happening in the way of motion of the puck? What happens to the net work done on the puck? What does the puck do if the force P suddenly drops to zero at position x?

(c) Show that for the more general case in which the force P is applied between an initial position x_1 and a final position x_2, the work–kinetic energy theorem requires that

$$\int_{x_1}^{x_2} P(x)dx - \left[(1/2)kx_2^2 - (1/2)kx_1^2\right] = (1/2)\,mv_2^2 - (1/2)\,mv_1^2 \qquad (2)$$

(d) Interpret eq. (2). What is happening if the left-hand side of the equation is positive? If the left-hand side is negative?

(e) Show that if the puck is released from rest ($v_1 = 0$) at some initial position x_1 not at the origin and with the force $P = 0$, eq. (2) requires that from then on

$$(1/2)kx_1^2 = (1/2)kx_2^2 + (1/2)mv_2^2 \qquad (3)$$

(f) Interpret eq. (3) and contrast the variation of the kinetic energy of the puck on the spring (when released from rest) with the variation of the kinetic energy of the ball released from rest in free fall. Show that very similar conservation relationships obtain, but that there are important differences in the character of the motion that can take place. If we drop a body vertically, from rest, through a displacement Δy, it acquires a kinetic energy $(1/2)mv^2$ equal to the magnitude of $mg\Delta y$ and will continue dropping and acquiring still more $(1/2)mv^2$ if the fall is not interrupted. If in the case of the spring, however, we release the puck from rest at a position x_1, it will be accelerated toward the left, and, on returning to $x = 0$, it will have acquired an amount of kinetic energy equal to $(1/2)kx_1^2$. It will then continue moving to the left, but the acceleration will now be directed to the right because of the stretching of the spring. In the case of free fall, there was no such reversal of the acceleration. Show that the equations tell us that the puck will continue moving to the position $x = -x_1$, at which point the direction of motion will be reversed. In the absence of friction, this oscillation would continue indefinitely (as in the case of the bouncing ball) with the maximum values of $(1/2)kx_1^2$ and $(1/2)mv^2$ being continually interchanged. The motion is "symmetrical" around the position $x = 0$. What mathematical characteristic of eq. (3) accounts for this symmetry? Solve eq. (3) for v_2 and interpret the result very carefully in words.

4.29 Look around the room or place in which you happen to be located. Identify and describe qualitatively at least two or three processes of transformation of energy that are going on right now in your vicinity. Do not ignore the fact that your own body may be involved in some of the processes.

Your description should be detailed in the sense of defining and describing relevant systems, changes in state, interactions, forms of energy involved, and forms of energy transfer that are taking place. Note that there is no possibility whatsoever that *no* energy transformations are taking place. We ourselves and everything around us are immeresed in a ceaseless flux of energy transformation.

4.30 When you are at the top of a staircase, a certain amount of gravitational potential energy is stored in the system consisting of your body and the earth. Describe what happens to this potential energy as you descend the staircase. Be sure to indicate clearly what other objects or systems become involved in the associated interactions and energy transfers and what forms of energy are involved.

4.31 The energy transformations taking place in chemical reactions (e.g., the amount of heat received from or transferred to surrounding systems) depend quantitatively on the initial and final states of the objects or materials constituting the reacting system. The initial state is defined by many properties or factors such as temperature, pressure, composition, internal stresses and their attendant deformations, and the effects of imposed electric and magnetic fields.

   (a) In the light of the foregoing statement, speculate on what happens to the stored potential energy when a *compressed* metallic spring is dissolved in acid and seems to disappear completely as far as its initial state, appearance, and configuration are concerned. Has the stored potential energy been ''destroyed,'' thus violating the conservation law? In what way is it likely that energy is still being conserved in these circumstances? (Compare this situation with what might be happening when the same *uncompressed* spring is dissolved in acid.)

Note to the student: In questions 4.32 and 4.33, circle the letters marking all those statements that are correct. Any number of statements may be correct, and therefore each one must be examined on its merits. Do not simply drop the question after you have found one correct statement.

4.32 A block of mass m is pulled along the floor by a force T inclined at an angle θ as shown in the following diagram. The coefficient of friction between the block and the floor is denoted by μ. The magnitude of the force is such that the block moves with uniform velocity.

(a) The magnitude of the normal force exerted <u>on</u> the block <u>by</u> the floor is given by:

(A) mg

(D) $mg - T\sin\theta$

(B) $mg - T\cos\theta$

(E) $T\sin\theta$

(C) $mg + T\cos\theta$

(F) None of the above.

(b) The magnitude of the frictional force exerted <u>on</u> the block <u>by</u> the floor is given by:

(A) $T\cos\theta$

(D) $\mu(mg - T\sin\theta)$

(B) $T\sin\theta$

(E) zero

(C) μmg

(F) None of the above.

(c) During a horizontal displacement Δx of the block:

(A) the work done by the force T is given by $T\Delta x$.

(B) the work done by the force T is given by $(T\sin\theta)\Delta x$.

(C) Some of the work done by the force T is stored as potential energy in the system while the rest is converted into thermal internal energy.

(D) the work done by the force T is equal to the kinetic energy change of the block.

(E) the change in potential energy of the block–earth system is $mg\Delta x$.

(F) None of the above.

4.33 A frictionless puck of mass m, mounted between identical springs as shown, can slide back and forth on the level frictionless surface. The springs have negligible mass relative to the mass of the puck.

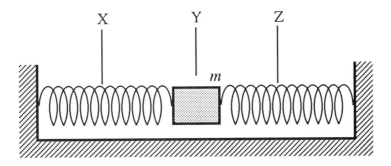

The puck is displaced by hand from its equilibrium position at Y to position X, at which point it is released from rest. It then oscillates back and forth between positions X and Z. Circle the letters marking the correct statements about the oscillatory motion.

(a) The puck has its largest value of kinetic energy at position Z.

(b) The puck has its largest value of kinetic energy at position Y.

(c) The system has its largest potential energy when the puck is at position Y.

(d) The system has its largest potential energy when the puck is at position Z.

(e) The potential energy of the system when the puck is at position Z is equal to the work that was done in displacing the puck from Y to X.

(f) The direction of momentum <u>change</u> of the puck is toward the right throughout any interval in which the puck is located between Y and X regardless of which way it is moving.

(g) The rate of change of momentum at position Y is zero.

(h) The rate of change of momentum has its largest <u>magnitude</u> at positions Y and Z.

(i) At *any* instantaneous position of the puck, the sum of the instantaneous values of kinetic energy of the puck and potential energy of the system is equal to the work that was initially done in displacing the puck from Y to X.

(j) None of the above.

Note to the student: Questions 4.34 through 4.36 describe a physical situation and then make a statement (or statements) about it. Accept the description of the physical situation (the numbered item) as given and correct. Examine the statements, including numerical values, to determine whether they are correct. If they are correct, say so explicitly. If they are incorrect, alter them to eliminate errors.

4.34 A frictionless puck on a string on a level air table moves in a horizontal circle at uniform angular velocity.

> (a) The string exerts a force on the puck, and the work done by this force is equal to the rotational kinetic energy of the bob.

> (b) The linear momentum of the puck does not vary as the puck continues in uniform circular motion.

> (c) The force exerted by the string keeps imparting an impulse of constant magnitude to the puck.

> (d) The vertical force exerted by the table on the puck delivers zero impulse to the puck.

> (e) The potential energy of the system consisting of the puck and the string can be taken to be zero.

> (f) The system consisting of the puck, the string, and the table can be regarded as a closed system whether or not friction is present.

4.35 A ball with a mass of 250 g is thrown vertically upward. It rises to a position 10 m above the point at which it left the thrower's hand. Neglecting frictional effects:

> (a) The velocity with which the ball left the thrower's hand must have been about 14 m/s.

> (b) The upward impulse delivered to the ball in the act of throwing must have been about 1.4 N·s.

> (c) The downward impulse delivered to the earth in the act of throwing must have been zero.

> (d)The kinetic energy imparted to the ball must have been about 2.5 J.

> (e) The work done by the thrower must have been about 2.5 J.

4.36 Two identical gliders move toward each other with equal speeds on a level air track.

(a) The total momentum of the system consisting of the two gliders is zero.

(b) The total kinetic energy of the system consisting of the two gliders is greater than zero.

(c) The kinetic energy of the system consisting of the two gliders will be reversed after the gliders have undergone perfectly elastic collision.

(d) With no external horizontal forces acting on the system consisting of the two gliders, the net impulse delivered to each one of the gliders must be zero over the interval of collision.

CHAPTER 5

Electricity

5.1 What is meant by the term "electrostatic interaction"? Describe the circumstances to which this name is applied. We also speak of "magnetic interaction" and "gravitational interaction." How do we distinguish one of these interactions from another? What are the similarities and what are the differences? (In answering this question, it is necessary to describe specific instances of what does and does not happen in these various phenomena. The larger your list the better the answer.)

5.2 (a) What is meant by the term "electrical charge"? Describe observed effects that lead us to invent this property for objects that have been treated in an appropriate way, even though we have no idea of what charge "is." (b) Describe observations that lead us to infer that charge, whatever it might be or however it might be carried, is movable or transportable from one object to another. (c) Describe observed effects that are interpretable in terms of the concept of altering the "quantity of electrical charge." (d) Describe observed effects that lead to the inference that the *strength* of electrical interaction (i.e., the force exerted by charged objects on each other) depends on at least two variables: separation between the objects and quantity of charge carried by each.

5.3 What is meant by the term "like" when we talk about electrical charges? How does this term originate? Answer the question by describing experiments and what is actually observed. What is the origin of the statement "like charges repel"? What is "alike" about like charges: Do they look, feel, sound, smell similar? Is the introduction of the word "like" a matter of arbitrary definition? Could we have equally well started with the definition "unlike charges repel" and based our system on that terminology, or do observations *dictate* the familiar terminology?

5.4 On what basis do we find that we can get away with only the two terms "like" and "unlike"? In other words, what is the basis for believing that no more than two varieties of electrical charge exist in nature even though we observe electrical interaction among a vast variety of different materials and in different circumstances? To make your argument convincing, you will need to describe observations or experiences (visualized in the abstract) that would force you to recognize and accept a third variety of electrical charge if it happened to turn up on a new material. The fact that no such interaction has ever been observed leads to the deeply held belief that there are no more than two varieties of charge. (Note that understanding the significance of what does *not* happen is sometimes just as important as knowing what does happen.) Where do the terms "positive" and "negative" come from? Are they necessary? Would other names do just as well?

5.5 What observed effects lead to the inference that both varieties of electrical charge are initially present in all objects in balanced amounts? What observed effects lead to invention of the model to which we give the name "polarization," visualizing the shifting or displacement of charges relative to one another in one object when another charged object is brought near.

5.6 Describe observations that lead to the invention of the concepts "conductor" and "nonconductor" or "insulator."

5.7 A hard rubber or plastic rod is rubbed with fur and is touched to an electroscope. The electroscope exhibits the presence of a net charge with the flexible leaf standing away from the fixed center post. If one brings one's hand or any large uncharged object near the electroscope platform, the leaf is observed to drop somewhat.

(a) Interpret this decrease in deflection by describing what we visualize as happening in the way of charge displacements in both the electroscope and the uncharged body. Do this by drawing diagrams showing the altered charge distributions and explaining your reasoning. How do you account for the fact that the shifting of charge does not simply go on indefinitely but that equilibrium configurations are attained?

5.8 A toy balloon, after being rubbed with a cloth or sheet of plastic, is brought up to the wall of the room and sticks to the wall without being held or otherwise supported. Explain in detail what is happening: Sketch the electrical effects that arise (not just in the balloon but also in the wall). Draw well-separated force diagrams for the balloon and for the region of the wall it touches while it is sticking to the wall. Describe each force in words and identify the

Third Law pairs. What holds the balloon vertically? Would the behavior of the balloon be different if it were hard and round instead of soft and deformable? If so, how would it behave?

5.9 Suppose we have four conductor-coated pith balls A, B, C, and D suspended on nonconducting (e.g., nylon) threads. The charge states of all four balls are initially unknown. We now take a rubber rod, rub it with fur so that it becomes negatively charged, and bring it in contact with sphere A. Then we bring the balls near each other (without contact) two by two. The interactions are as follows: (1) B, C, and D are all attracted to A; (2) B and C have no discernible effect on each other; (3) B and C are both attracted to D.

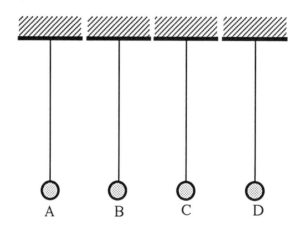

(a) What are the charge states of A, B, C, and D? Explain your reasoning.

(b) In the story above and its conclusions, what were the observations and what were the inferences?

Suppose, by handling sphere C with the nylon thread, we now bring it close to (but not touching) sphere A. While the two spheres are close together, we touch C briefly with one finger. Then we remove C from the vicinity of A.

(c) What is now the charge state of C? Explain your reasoning, describing, with words and pictures, what happens to C step by step through the entire sequence. What is the name for the phenomenon you have just described?

(d) How will C now interact with A, B, and D? Explain your reasoning.

5.10 Two identical conducting spheres A and B carry equal amounts q of like charge and, separated by a distance that is large compared to their diameters, repel each other with a force of magnitude F. A third identical conducting sphere D is mounted on an insulating handle and can be moved around at will by the experimenter. The experimenter first brings the initially uncharged sphere D in contact with sphere A, and then, without discharging D, brings it in contact with sphere B. Sphere D is then removed from the vicinity of A and B.

(a) How will the magnitude of the force with which A and B now repel each other compare with the initial value F, the distance between A and B remaining unaltered? Give a numerical value for the ratio and explain your reasoning.

(b) How will the amount of charge now carried by D compare with the initial value q? Give a numerical value for the ratio and explain your reasoning. (Be sure to verify that charge has been conserved.)

5.11 Two metal cans are placed near each other on a table as shown. A pith ball on a nylon string is suspended so that it hangs between the two cans. One of the cans is now charged positively by contact with a positively charged rod. The pith ball is attracted to the charged can, makes contact, swings over to the other can, makes contact, flies off, and then continues to oscillate back and forth between the two cans.

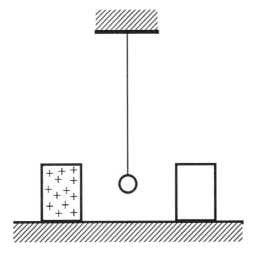

(a) Sketch a series of diagrams in which you show (1) how the uncharged pith ball initially becomes attracted to the charged can; (2) what happens when it makes contact with the charged can; (3) why it swings over to the other can and what happens when it makes contact there; and (4) why it continues to swing back and forth. Accompany each of your diagrams with an explanation of what is happening. You should be able to carry out this description in any one of the following ways: (1) conventional displacement of positive charge with negative charge fixed; (2) displacement of negative charge with positive charge fixed; or (3) both varieties of charge displaced.)

(b) How long will the back-and-forth oscillation continue? What is the criterion for its stopping?

5.12 Metal spheres A and B, standing on insulating supports, are in contact with each other. Another sphere C, highly charged negatively through contact with a Van de Graaf generator or a Wimshurst machine, is brought near A and B as shown. While C is nearby, B is moved off to the right so that A and B are now separated. C is then removed from the vicinity.

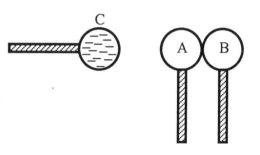

(a) A and B are now charged. Describe, with the help of appropriate diagrams, how they became charged, and identify the kind of charge present on each. What is the name of the process by which B became charged?

When B is now brought back near A (without contact), a spark is likely to jump between the spheres with an obvious appearance of light, heat, and sound.

(b) How did it become possible for energy to be released in this manner? Describe the energy transformations that took place in the entire sequence outlined.

5.13 In the sequence illustrated in the following diagram, A shows an electroscope carrying a net positive charge and exhibiting a deflection of its needle accordingly. (Remember that in the convention we use to indicate charging, we show only the *excess* variety of charge in various regions, not the underlying internal sea of uniformly distributed positives and negatives.)

A negatively charged rod is now brought toward the electroscope in three successive steps, somewhat closer in each step, as illustrated. In B the electroscope shows somewhat less deflection than in A. In C the deflection is just zero. In D the deflection is again large.

(a) In each of the diagrams B, C, and D, sketch an arrangement of excess positive or negative charges in various regions to show what happens on the electroscope as the charged rod is brought closer and the deflection of the needle changes.

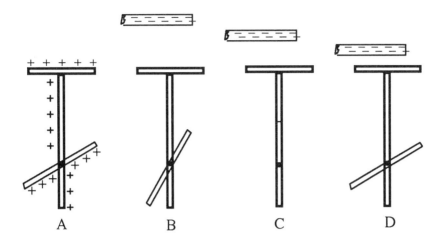

A   B   C   D

5.14 An uncharged metal rod B rests on a glass beaker as shown. A conductor-covered pith ball C hangs on a nylon thread near the right-hand end of the rod. When a strongly negatively charged plastic rod A is brought near the left-hand end of B (without making contact with B), the pith ball is attracted to the rod, makes contact with the rod, and then flies off. The plastic rod is then removed, still without having touched B.

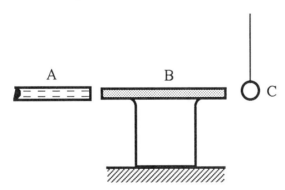

(a) With an appropriate sequence of sketches of your own choice and with accompanying verbal description, show what happens in this system, stage by stage, with respect to charging and charge distributions, and predict the final charge condition of both rod B and pith ball C.

5.15 Consider the situation in which a hollow conducting sphere A is located *inside* another hollow conducting sphere B. Sphere B has twice the diameter of A. The two spheres are connected by a metal wire as shown. A quantity of charge Q is transferred to sphere B by contact from an electrostatic generator.

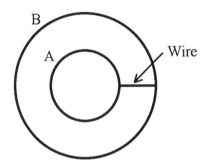

(a) What fraction of the charge Q is transferred to sphere A as equilibrium is attained in the system? Explain your reasoning.

5.16 Given the fact that two charged metallic conductors are in equilibrium with respect to transfer of charge from one to the other when their potentials relative to infinity are equal. Given also the fact that the potential of an isolated, charged metal sphere is inversely proportional to its radius R.

Now consider the case in which an uncharged sphere A with radius R_A is allowed to come to equilibrium with a charged sphere B with radius R_B, carrying a quantity of charge Q. (Since bringing the spheres into direct contact, surface to surface, results in drastic alteration of the charge distribution and may do complicated things to the potential, let us perform the "thought experiment" of achieving equilibrium by keeping the two spheres fairly widely separated and transporting charge from one to the other by moving a very small conducting sphere back and forth between them until no more charge is transferred. Under these circumstances, the two principal spheres maintain reasonably well-defined potentials, and equilibrium is attained when the two potentials are equal. (Still another thought experiment approximating the same situation might be the connecting of the two spheres by a long, thin wire.)

(a) Explaining your reasoning, obtain expressions for the final quantities of charge q_A and q_B carried by each of the spheres at equilibrium. Give the expressions in terms of the radii and Q, the initial total charge on sphere B.

(b) Interpret the algebraic expressions you have obtained. How is the charge distributed when the two spheres are identical? What happens to the charge received by A as R_A is made very much smaller than R_B? As R_A is made very much larger than R_B?

(c) Consider the following assertion: "When R_A is very small relative to R_B, the charge received by sphere A is actually a close measure of the initial potential of sphere B; when R_A is very much larger than R_B, the charge received by A becomes a measure of the quantity Q initially carried by B." Is this assertion correct or incorrect? Explain your reasoning.

(d) To a crude approximation, an ordinary electroscope might be treated in terms of the ideas developed above. In the light of the conclusions reached in part (c), what would you say the deflection of an electroscope leaf measures (approximately) when the electroscope makes contact with an object very small relative to itself? When the electroscope makes contact with an object very large relative to itself?

(e) Suppose the initially charged hollow sphere B is much larger than sphere A and has a hole in its surface so that objects on handles can be inserted inside the spherical shell. Uncharged sphere A, carried on an insulating handle, is inserted into sphere B and brought in contact with the inside surface of B. The initial charge of B is still Q, and the radii are R_A and R_B, respectively. What is the final equilibrium distribution of charge between A and B under these circumstances? Explain your reasoning both in terms of distribution of charge on the hollow sphere and the concept of potential, showing internal consistency among the lines of reasoning.

5.17 We have a pair of large capacitor plates oriented and charged as shown in the following diagram. We also have a particle of mass m, carrying charge $+q$ so small that its presence has negligible effect on the charge distribution on the plates. No gravitational or other forces are to be considered. The potential difference between the plates is denoted by ΔV, the spacing between the plates by Δs, and the magnitude of the electrical field strength between the plates by E.

(a) Suppose that our test particle is placed successively at positions A, B, and C. Show, on a force diagram for the particle, the force acting on the particle at each one of these positions. If the force happens to be zero, say so explicitly. Explain your reasoning.

(b) If the test particle is displaced from position D to position E, what amount of work must be done on (or taken out of) the system? Explain your reasoning.

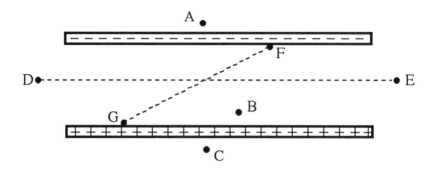

(c) If the test particle is displaced from position F to position G, what amount of work must be done on (or taken out of) the system? Explain your reasoning.

5.18 Two positively charged conducting spheres A and B are located with their centers 10.0 cm apart. The two charges are *unequal*, A carrying 2.5×10^{-8} C and B carrying 1.2×10^{-8} C. Draw a force diagram for each sphere, assuming the presence of fixed supports that keep the spheres from accelerating. Calculate the force exerted by A on B and the force exerted by B on A. Explain your reasoning.

5.19 The charged particles A, B, and C, occupy fixed positions at the vertices of a right triangle, as shown. The charges on the particles are all equal in magnitude. No interactions other than electrostatic are present.

(a) Draw a force diagram for each particle showing all the forces acting on it. Describe each force in words. Identify the Third Law pairs. (Show larger forces with longer arrows and smaller forces with shorter arrows.)

5.20 The following diagram shows the region in the neighborhood of a large positively charged conducting plate (extending far beyond the region shown) and a negatively charged conducting sphere. Lines labeled A – F are shown and are said to be field lines for the electric field between the two charged objects.

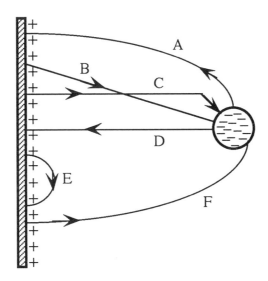

(a) Examine each one of the lines, and indicate whether it is a correctly drawn field line. If any line is not correct in some way, explain what is incorrect about it.

(b) Redraw the diagram to make the pattern of field lines more nearly correct.

5.21 Four particles, each carrying the same magnitude of electrical charge, are located at the corners of a square as shown. Point A is located at the midpoint of one side of the square; point B is located precisely at the center.

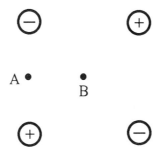

(a) Determine *qualitatively* (numbers are not needed) whether there is a nonzero electrical field at each point (A and B). If the field is not zero, indicate its direction. Show how you came to your conclusions.

(b) Let us return to the situation at point B (no gravitational effects being present). Suppose you placed a charged particle at B. Would it stay there? Why or why not? (Compare this situation with trying to balance a ball bearing on the very top of a bowling ball. Such points are called positions of "unstable equilibrium.")

5.22 Two charged particles are located along the x-axis, as shown. We define region A as that for which $x < 0$, region B as that for which $0 \le x \le x_1$, and region C as that for which $x > x_1$.

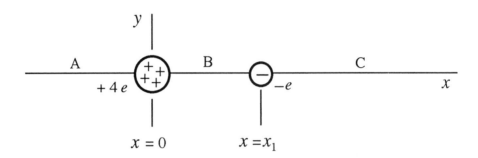

(a) The electrical field strength along the x-axis is, of course, zero at both $+\infty$ and $-\infty$. Is the electrical field strength zero at any other point (or points) along the x-axis ? Explain your reasoning.

(b) Identify the precise location of the point at which the field strength is zero. (The calculation does *not* require an algebraic analysis; it can easily be done mentally.)

(c) Now consider a more general case: suppose the two particles, located as in the figure, might carry any reasonable quantity of charge of either variety. Are there cases in which there is no point, other than at infinity, where the electrical field is zero along the x-axis? If so, describe them. Are there cases in which the electrical field will be zero at more than one point (other than at infinity) along the x-axis? If so describe them.

5.23 A positively charged ball is suspended on a nylon (non-conducting) thread from a positively charged metal plate of very large extent, as shown in the following diagram. The ball swings back and forth as a pendulum bob. (Both electrical and gravitational effects are present.)

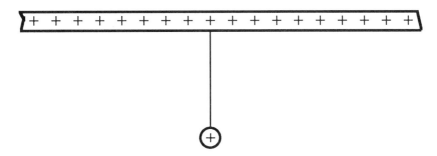

(a) Draw separate force diagrams for the ball and the thread at some arbitrary angle of deflection from the vertical, and describe each force in words. How will the tension in the string in the presence of electrical charge compare with the tension that would obtain in the presence of gravity alone? Explain your reasoning.

(b) How will the period of the swing in the presence of electrical charge compare with the period in the presence of gravity alone: Will it be equal to, greater than, or less than the latter? Explain your reasoning.

5.24 Suppose an irregularly shaped three-dimensional region such as that sketched here contains $+3.6 \times 10^{-8}$ C of positive charge uniformly distributed throughout the region. There is no charge present outside the region indicated. Suppose we now scale the three-dimensional region *down*, shrinking each of the x, y, and z dimensions by a factor of 1.85 while keeping all the original charge within the smaller region.

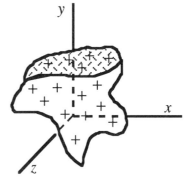

(a) How would the final charge *density* τ_2 compare with the initial charge density τ_1? Calculate the numerical value of the ratio and explain your reasoning. (The term "charge density" refers to amount of charge per unit volume.)

(b) As we shrink the region down uniformly from all sides, what will happen to the force exerted on a spherical charge of -4.8×10^{-7} C which remains at a fixed position at a distance very large compared with the dimensions of the positively charged region we are shrinking, i.e., will the force on the distant negative charge increase, decrease, or remain essentially unchanged? Explain your reasoning.

5.24 Point P is located at a distance d from a negatively charged metal plate that has dimensions very much larger than d. A small positively charged metal sphere, carrying charge q, is located near P at the same distance d from the plate.

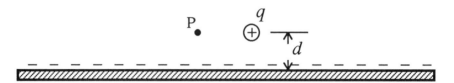

(a) If point P and the metal sphere are both moved up to distance $2d$ from the plate, how will the electrical field strength at the new location of P compare with the field strength at the initial location? Will it be increased, decreased, or essentially the same? Explain your reasoning.

5.26 Recall the phenomena associated (1) with frictional effects, such as rubbing various objects together; (2) with chemical effects, such as those taking place in (voltaic) batteries and in electrolysis; and (3) with effects taking place in so-called "electromagnetic generators." It is asserted that these are all manifestations of electrical effects involving transfer of electrical charge. In other words, all these seemingly very different phenomena are intimately related.

(a) Suppose you are confronted by a well-educated nonscientist friend who is, very reasonably, skeptical of the assertion, being conscious of the immense disparity among the various phenomena and not believing them to be linked to the same basic concept. How would you proceed to convince him or her that the assertion is indeed correct? Your argument should include appeal to experiments and demonstrations that might be carried out, not just verbal argument. (Faraday devoted an entire, important paper to listing and describing just such demonstrations proving the relatedness of the phenomena.)

5.27 An electrical engineer files a patent claim on an arrangement of charged electrodes specially shaped so that, in a *charge-free* cylindrical region the axis of which is parallel to the x-axis, there exists an electrical field with lines of force precisely parallel to the x-axis and with magnitude of field strength increasing linearly according to the equation $E = bx + d$. The equation applies in the region $0 \leq x \leq a$, and b and d are constants.

(a) Sketch a diagram of the situation that has been described.

(b) If you were the patent examiner, would you approve or disapprove the claim? Explain your reasoning. (If you have studied Gauss's law, present your argument by application of this law.)

Note to the student : In the following multiple-choice questions, circle the letters designating those statements that are true or correct. <u>Any number</u> of statements, <u>not</u> just one, may be correct. You must consider each statement on its merits and not simply stop when you have found one correct statement.

5.28 A positively charged particle carrying 2.0×10^{-8} C enters a region between charged capacitor plates through a hole in one plate, as shown. The potential difference between the plates is 1000 V, and the kinetic energy of the particle as it enters the hole is 1.0×10^{-5} J. (Only electrical effects are to be considered. Gravitational effects and air resistance are to be ignored.)

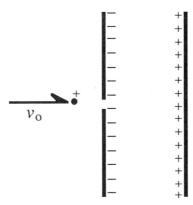

(a) The kinetic energy of the particle remains unchanged as it moves toward the right-hand plate.

(b) The kinetic energy of the particle decreases as it moves toward the right-hand plate.

(c) The particle has insufficient kinetic energy to reach the right-hand plate, and it "falls back" toward the hole after going part way.

(d) The particle collides with the right-hand plate and bounces back toward the left-hand one.

(e) As the particle moves toward the right-hand plate, the potential energy of the particle–capacitor system increases.

(f) The data given are insufficient to allow calculation of the force acting on the particle when it is between the plates.

(g) The momentum of the particle is conserved throughout its motion between the plates.

(h) None of the above.

5.29 A capacitor is first charged by connecting it, by means of wires and a switch, to a battery having a potential difference of 120 V. The switch is then opened, and the capacitor remains charged. What happens when a metal wire is inserted, connecting the charged plates?

(a) A spark jumps from one plate to the other.

(b) The charge on the plates is reversed.

(c) The charge on both plates becomes zero.

(d) The potential difference between the plates becomes zero.

(e) The electrical field strength between the plates is reversed.

(f) The electrical field strength between the plates drops to zero.

(g) The kinetic energy of the system is converted to potential energy.

(h) The capacitance becomes zero.

(j) None of the above.

5.30 A particle carrying positive charge q_A is subjected to the superposition of two electrical fields — one due to the charged capacitor plates and the other due to the negatively charged particle q_B as shown the following diagram. The magnitude of charge q_B and its distance from q_A are such that the electrical field strength due to q_B at the position of the positively charged particle is equal in magnitude to the field strength due to the charged capacitor.

Which arrow in the diagram best shows the direction of instantaneous acceleration of the positively charged particle?

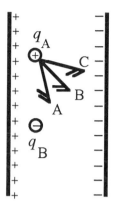

(a) Arrow A.

(b) Arrow B.

(c) Arrow C.

(d) Any of the above, depending on the instantaneous velocity of the positively charged particle in its given location.

(e) None of the above, since the two fields cancel each other.

CHAPTER 6

Direct Current Circuits

6.1 You are given a flashlight battery, a flashlight bulb, and a *single* flexible wire as shown.

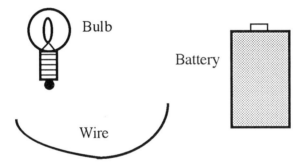

(a) Sketch at least two or three different arrangements of these three items that would result in lighting the bulb.

(b) Sketch at least two or three different arrangements that would *not* result in lighting the bulb.

Note to the instructor: Many students have trouble translating a circuit layout on the laboratory table into a conventional circuit diagram and vice versa. Physical as well as pencil-and-paper exercises in such translation, in *both* directions, are well worth posing for homework and on tests in order to help generate control of this mode of thinking. Complex configurations are not to the point. Simple configurations, such as those illustrated in some of the following problems, are fully adequate to this purpose.

6.2 In these three circuits, identical bulbs, in different combinations, are lighted by connection to identical, ideal batteries. (Ideal batteries have zero internal resistance and suffer no drop in potential difference across their terminals when an external load is connected.)

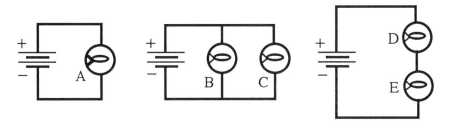

(a) Rank the five bulbs (A, B, C, D, E) in order of decreasing brightness, indicating equal brightness when such is the case. Explain your reasoning.

(b) Rank the three circuits in order of increasing current drawn from the battery. Explain your reasoning.

(c) How does the situation in the following diagram compare with that in part (a)? Is it the same or different? If there are differences, describe them. Be sure to explain your reasoning.

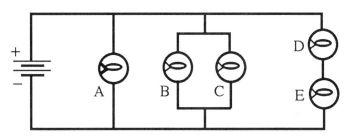

6.3 In the circuit shown in the following diagram, the battery maintains a constant potential difference across its terminals as various changes are made in the circuit containing the three identical bulbs (A, B, and C). You will be asked to predict what will happen as the various changes are made.

(a) To begin with: How do the brightnesses of the three bulbs compare with each other in the initial condition as sketched? Explain your reasoning.

(b) Suppose bulb C is removed from its socket. Will the brightnesses of bulbs A and B change? If so how? What will happen to the current at point 3? Explain your reasoning

(c) Return to the initial condition in the diagram. Suppose a wire is connected between points 3 and 4 in the circuit. What will happen to the brightness of each bulb? What will happen to the current at point 2? What will happen to the potential difference between points 2 and 4? To the potential difference between points 4 and 5? What will happen to the current at point 5? Explain your reasoning.

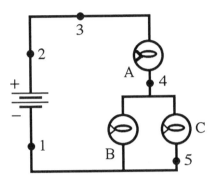

(d) Return to the initial condition. Suppose a wire is connected between points 4 and 5. Answer the same questions as those asked in part (c). Explain your reasoning.

(e) Return to the initial condition. Suppose a third bulb, D, is added to the circuit by being placed in parallel with B and C. Answer the same questions asked in part (c). Explain your reasoning.

(f) Return to the initial condition in the diagram. Suppose a wire is connected between points 1 and 5 in the circuit. What will happen to the brightness of each bulb? To the current at point 3? To the potential difference between points 3 and 4? Explain your reasoning.

(g) Make up a configuration of your own and investigate it by asking questions and making predictions of the variety illustrated above, together with any additional questions you can invent for yourself. One of the best things you can do to strengthen your understanding of simple electric circuits is to obtain or borrow some batteries, bulbs, sockets, and wire and test your own predictions on behavior in various configurations. (The kind of thinking you are asked to do in this question is exactly the kind of thinking that is used in troubleshooting virtually any electric circuit.)

6.4 The following circuit contains two identical flashlight bulbs A and B and three identical resistors R. The battery maintains a constant potential difference across the circuit regardless of the various changes proposed in the questions. Be sure to explain your reasoning in each case.

(a) How do the brightnesses of bulbs A and B compare initially?

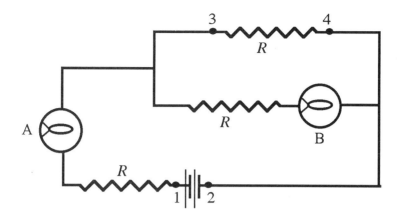

(b) Bulb A is removed. What happens to the brightness of B?

(c) Bulb A is replaced, and then B is removed. What happens to the brightness of A?

(d) Bulb B is replaced. A wire is connected from point 1 to point 3. What happens to the brightness of each bulb? What happens to the potential difference between points 3 and 2?

(e) Return to the initial situation. A wire is connected from point 3 to point 4. What happens to the brightness of each bulb? What happens to the potential difference between points 2 and 3? To the potential difference between points 1 and 3?

(f) Return to the initial condition. A wire is connected from point 2 to point 4. What happens to the brightness of each bulb and to the potential difference between points 1 and 4?

(g) Return to the initial condition. A fourth resistor, identical to the other three, is connected between points 3 and 4. What happens to the brightness of each bulb and to the potential difference between points 3 and 4?

Note to the instructor: It is obvious that any number of similar questions can be formed with different (simpler or more complex) configurations. For example, one can construct questions parallel to 6.3 and 6.4 in connection with the slightly simpler circuit that follows (left) or the slightly more complex circuit (right). In connection with the circuit on the right, one might additionally ask about the effect that opening or closing the switch S would have on the brightnesses of the bulbs. One might also add bulbs to an initial configuration.

It takes several exercises of this kind to bring a substantial number of students to the point of dealing with them correctly — even students who manage conventional end-of-chapter problems using Ohm's law and Kirchhoff's laws. After giving one or two examples, it is advantageous to challenge students to make up their own configurations and questions in a contest to outguess the instructor. Those students who accept the challenge rapidly improve their performance and exhibit higher morale as well as satisfaction in achievement.

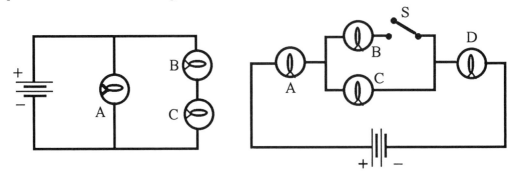

6.5 In the light of what you have observed happening with series and parallel electric circuits, what do you infer to be the arrangement utilized in your household electric system? Are lights, appliances, etc., connected in series or in parallel or in some combination thereof? Explain your reasoning and your available evidence.

6.6 This circuit consists of a flashlight battery, a flashlight bulb, and a filament of steel wool of grade 0 or 1 (wiggly line) inserted, as sketched, in series with the bulb. The bulb is lighted under these circumstances.

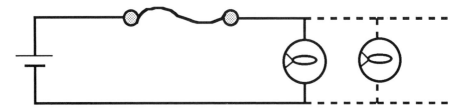

(a) Suppose you now proceed to add more and more additional bulbs in parallel with the first one as shown by the dashed lines. What do you expect might happen at some point to the filament of steel wool? (Try such a set-up yourself and see. The fact is that the filament burns out.)

(b) What does the observed burning out of the filament indicate regarding the size of the electric current drawn from the battery as the number of parallel bulbs is increased? Explain your reasoning.

(c) Explain the analogy between the situation described above and the role played by fuses or circuit breakers in your household electrical system. What happens as you plug more and more appliances into the same outlet (or into outlets on the same circuit)? Why are fuses or circuit breakers required by law?

(d) How would you expect the lifetime of a battery used to light one bulb to compare with the lifetime of a battery used to light two bulbs in parallel? Explain your reasoning.

6.7 Any conducting object has its own resistance to electric current. Suppose we insert more and more such objects in series across the terminals of a given battery.

(a) Describe the observations that indicate what happens to the current drawn from the battery as the objects are added. In the light of what happens to the current, what do you infer is happening to the *total* resistance connected across the battery? How does the *total* resistance compare with the resistance of any one of the objects in the system? Explain your reasoning.

(b) Suppose instead of adding the objects in series, we keep adding them in parallel. Describe observations that indicate what now happens to the total current drawn from the battery. In the light of what happens to the total current, how does the combined effective resistance of the group of objects in parallel compare with the individual resistance of any one of the objects? Explain your reasoning.

6.8 Consider all the experiences you have now had with electric circuits (including experiments, demonstrations, observations made at home, etc.). In all the combined experience at the *macroscopic* level, is there any identifiable evidence concerning the *direction* of flow of either positive or negative charge? ("Yes" or "no" is not an adequate answer. It is necessary to discuss observational evidence.)

6.9 Is it possible to connect two identical batteries to make two bulbs in series light just as brightly as one alone across one battery? If so, sketch and explain the circuit diagram. If not, explain your reasoning.

6.10 Suppose you have lighted a flashlight bulb by connecting it across a battery. While the bulb is lighted, you now connect a wire directly across the terminals of the bulb or its socket. What is observed to happen to the brightness of the bulb? How do you interpret the effect in terms of the concepts of "current" and "resistance"? In terms of our model of what happens in electric circuits, what do you infer happens to the current that passes through the bulb? Does it literally drop to zero or does it level off at some other value? Why is this arrangement unhealthy for the battery? Explain your reasoning.

6.11 You have certainly heard the term "short circuit" used in a variety of circumstances. Define this term *operationally*. (This means that you must describe actions and observations of results of these actions. Words or synonyms alone are not adequate.)

6.12 Examine a switch, a socket, and a light bulb if you have never yet done so. It is worth breaking an old light bulb (well wrapped in heavy cloth) to be able to see what is inside, but this should be done with great care.

(a) Do switches, sockets, and light bulbs have nonconducting material as well as conducting material present in their structure? If non conducting material is present, what role does it play? That is, what would happen if it were not there? As part of your explanation, sketch the basic structure of each one of these objects.

6.13 A current of magnitude I divides at the junction shown to pass through an ammeter having a resistance of 3.0 Ω and a resistor of 27.0 Ω.

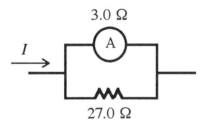

3.0 Ω

27.0 Ω

(a) What fraction of the current I passes through the meter? Explain your reasoning.

(b) What would happen to the magnitude of this fraction if the resistance of the resistor were increased? Explain your reasoning.

6.14 In the following circuit diagrams, the circle around the symbol A represents an ammeter; the circle around the symbol V represents a voltmeter. Consider these circuit diagrams in terms of what you have learned about the properties and function of ammeters and voltmeters.

(a) Compare the brightnesses of the five bulbs (B, C, D, E and F) with the circuits wired as indicated. Explain your reasoning.

(b) Explain why circuit E is not healthy for either the battery or the ammeter.

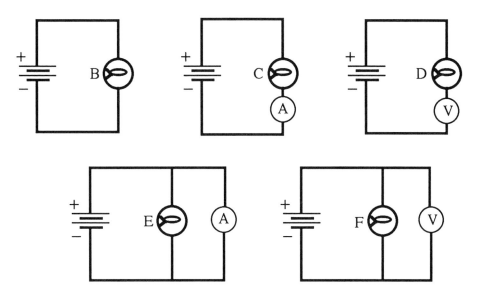

6.15 Suppose you are given two boxes that conceal what is inside, although each has two wires coming out of the interior. You have at your disposal some batteries, wire, and identical flashlight bulbs. Describe how you might go about determining which of the two boxes has the lower electrical resistance or whether the two are identical. Show the diagrams you would employ and explain your reasoning.

6.16 The ten circuits illustrated next contain an unknown resistance R and various arrangements of ammeters and voltmeters. Explain your reasoning in answering each of the following questions.

(a) With which circuit (or circuits) would the arrangement allow measurement of the value of the resistance R if the potential difference at the battery terminals is unknown?

(b) With which circuit (or circuits) would the arrangement allow measurement of the value of the resistance R if the potential difference at the battery terminals is known?

(c) In which circuit (or circuits) would the voltmeter read zero?

(d) In which circuit (or circuits) would the ammeter read zero?

(e) Which circuit (or circuits) would burn out the ammeter?

(f) Which circuit (or circuits) would burn out the voltmeter?

(g) Which circuit (or circuits) would allow determination of the power dissipated in R if the value of R is known?

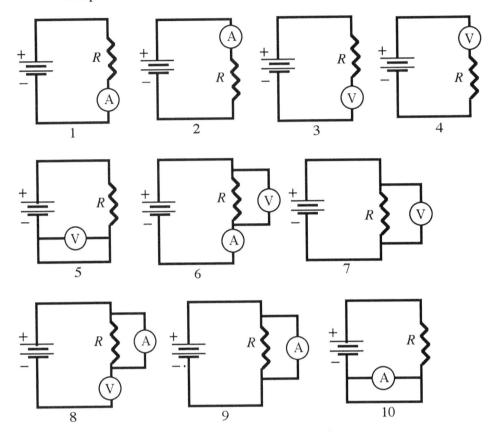

6.17 The following electrolysis circuits, I, II, and III, contain identical electrolytic cells connected to identical batteries. The oxygen is allowed to escape to the air while the hydrogen is collected in the inverted test tubes, which are initially completely filled with the electrolytic solution. All three circuits are run for the same length of time at room temperature and pressure, and hydrogen is collected.

(a) Examine the circuits and indicate any series arrangements or parallel arrangements.

(b) How would you expect the volumes of hydrogen in the various test tubes to compare with each other? Label the test tubes, and give their comparative rank, being sure to indicate equality of volumes when that is the case. Explain your reasoning.

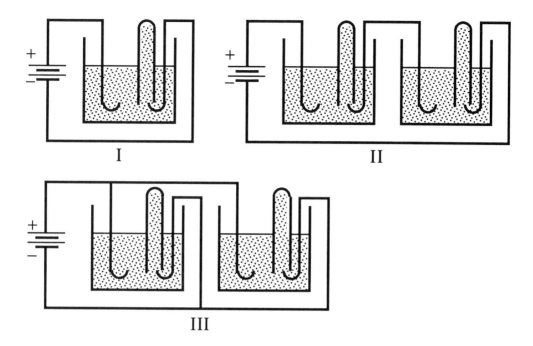

6.18 Two different light bulbs, A and B, are connected in parallel across the same battery. Bulb A lights more brightly than bulb B. In what property do the two bulbs differ? How do they differ? Explain your reasoning.

6.19 An *open* circuit contains a resistance R, an ammeter A, and a battery having emf ε and internal resistance r, as sketched in the following diagram.

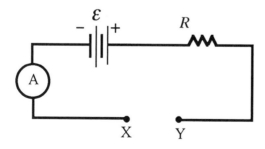

(a) Under these circumstances, what is the potential difference between points X and Y? Explain your reasoning.

6.20 Consider the following circuit in which the battery has an emf ε and internal resistance r. A voltmeter V and ammeter A are placed as shown.

(a) When the switch S is closed, how do the voltmeter and ammeter readings compare with the readings that obtained before S was closed? That is, do they increase, decrease, or remain unchanged? Explain your reasoning.

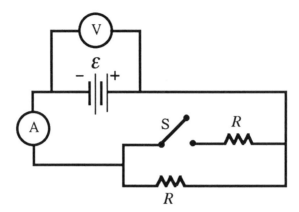

6.21 A wire of one metal has twice the resistivity, twice the diameter, and twice the length of a wire of another metal.

(a) What is ratio of the resistances of the two wires? Give a numerical value and explain your reasoning.

6.22 Two resistors are connected in parallel across a battery with negligible internal resistance, as shown. One of the resistors carries a current of 1.00 A while the other carries a current of 2.00 A.

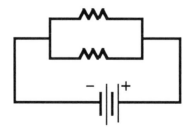

(a) Approximately what will be the current in the circuit if the two resistors are now connected in series instead of in parallel? Explain your reasoning.

6.23 A long, uniform resistive wire has a resistance such that, when it is connected to a battery as shown in circuit (a), the current in the circuit is 1.00 A. The circuit is now altered in such a way that the moveable contact (indicated by the arrow) is positioned at the center of the resistive wire rather than at the right-hand end as in (b), and the right-hand end of the wire is connected back to the left-hand end. Assume that the battery has negligible internal resistance.

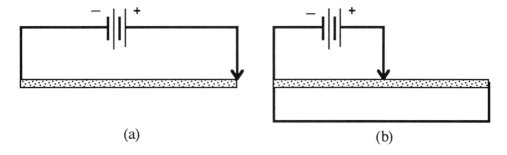

(a)   (b)

(a) What will be the current through the battery in circuit (b)? [You might find it helpful to redraw circuit (b) in a way that makes the pattern of resistances more familiar.] Explain your reasoning.

6.24 A circuit contains three identical bulbs lighted by connection to a single battery. When a wire is connected across the terminals of one of the bulbs, it goes out and so does another one of the bulbs. The remaining bulb becomes brighter.

(a) Sketch the circuit in which the bulbs must have been arranged. Explain your reasoning.

(b) What would happen to the brightnesses of the other two bulbs if the wire were connected across the terminals of the bulb that does not go out in part (a)?

6.25 Suppose we have two wires X and Y of different metals. The wires have the same resistance at room temperature, but the resistance of X increases more rapidly with increasing temperature than does that of Y. (If their temperatures are equal, the wires lose heat to the surroundings at the same rate.)

(a) If the wires are connected in parallel across a battery, which wire will have the higher temperature when the system reaches equilibrium? Explain your reasoning.

(b) If the wires are connected in series across a battery, which wire will have the higher temperature when the system reaches equilibrium? Explain your reasoning.

6.26 Three resistors are connected to a battery as shown. Let us suppose that the battery maintains a constant potential difference ΔV across its terminals and that the resistances differ in magnitude so that $R_1 > R_2 > R_3$.

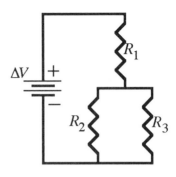

(a) Rank the current drawn from the battery and the currents in each of the three resistors in order of increasing magnitude. Explain your reasoning.

(b) Rank the potential difference across the battery terminals and the potential differences across each of the three resitors in order of increasing magnitude. Explain your reasoning.

(c) Describe in words the circumstances under which the potential difference across R_1 would be equal to that across R_2.

6.27 Four resistors, connected to a battery as shown in the following diagram, have their relative values indicated. The battery maintains a constant potential difference ΔV across its terminals.

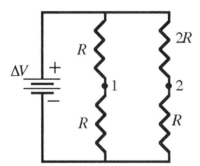

(a) If you connected a voltmeter between points 1 and 2, would it measure a finite or a zero potential difference? Explain your reasoning. If the potential difference is not zero, which point is more positive? (The upper terminal of the battery is the positive terminal.) Explain your reasoning.

(b) Suppose a wire is now connected between points 1 and 2. (This makes the potential difference between points 1 and 2 zero.) Will there now be a current in this wire? If so, in what direction (assuming conventional positive current)? Explain your reasoning.

(c) Suppose the upper right-hand resistor in the diagram had the same resistance R as the other three resistors instead of the value $2R$. Answer the same questions as in parts (a) and (b).

6.28 H is a resistive metal hoop to which a battery is connected at points P and Q. Contact Q is fixed while contact P can be moved at will around the hoop, varying the angle θ from 0 to 2π.

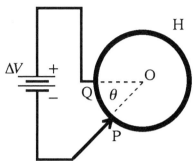

(a) What is the value of the resistance across the battery when P coincides with Q? Explain your reasoning.

(b) Does the resistance across the battery change as contact P is moved counterclockwise starting at Q? If so, does it increase or decrease? Explain your reasoning.

(c) Is there a position of contact P at which the resistance across the battery is greatest? If so, where? Explain your reasoning.

6.29 Consider these two circuits with identical batteries. In each case resistance R_1 is greater than R_2.

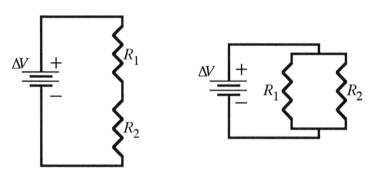

(a) In each circuit, which resistance dissipates the larger amount of power? Explain your reasoning.

(b) Explain how it comes about that the same resistance does not dissipate the larger amount of power in each of the circuits.

(c) In discussions of power P dissipated in a resistor obeying Ohm's law, it is shown that two relations are possible: (1) $P = I^2 R$ and (2) $P = (\Delta V)^2 / R$. Are they both applicable to any situation that arises?

Why or why not? What relevance does this question have to your answers in parts (a) and (b)?

6.30 The graph shows a record of current in amperes versus time in minutes. The current was drawn from a 120 V direct current source to operate some direct current motors.

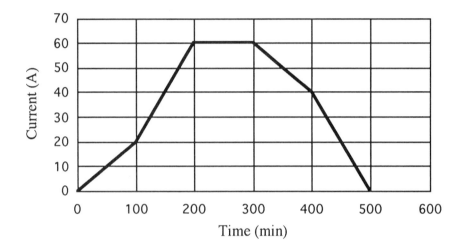

(a) If the cost of electric power under these circumstances is 5.0 cents per kilowatt-hour, calculate the total cost in dollars of the electrical energy supplied. Show and explain all steps of your calculation.

Note to the student: In the following multiple choice questions, circle the letter or letters designating those statements that are <u>correct.</u> <u>Any number</u> of statements may be correct, <u>not</u> just one, and you must examine each statement on its merits.

6.31 Suppose you are given an ideal battery that maintains a constant potential difference ΔV across its terminals regardless of which of the following resistive loads is connected to it. You are also given three identical resistors, each with a resistance of R ohms. You wish to heat water in a beaker with a setup such as that sketched in the following diagram.

(a) Which of the following combinations of the resistors would you put into the box with the question mark to produce the most rapid heating of the water?

(b) Explain your reasoning, and, if you choose (G), indicate what other
combination you would introduce and why.

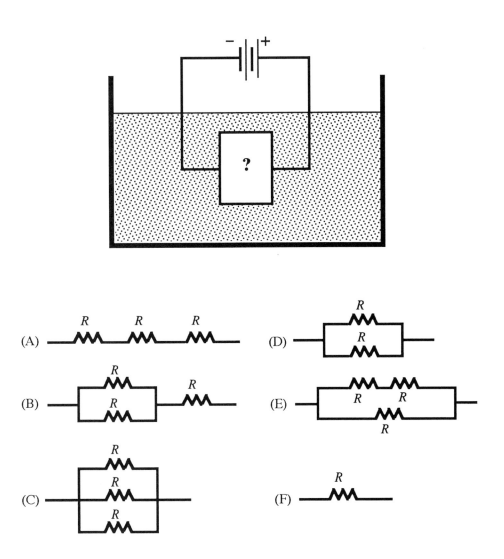

(G) NONE OF THE ABOVE.

Note to the instructor: One can alter question 6.31 to demand more courage
of conviction by leaving out configuration C and making the correct answer G,
asking students to sketch the correct combination if G is chosen.

6.32 A capacitor is connected to a 120 V battery as shown in the following diagram. A very high resistance wire is then inserted, connecting points A and B on the plates.

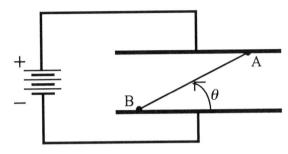

The potential difference between the ends of the wire will then be

(a) 120 $\cos\theta$ V

(d) 120 $\sin\theta$ V

(b) 120 V

(e) 120/$\cos\theta$ V

(c) 120/$\sin\theta$ V

(f) $\tan\theta$ V

(g) None of the above; once the wire has been inserted, the potential difference cannot be determined.

6.33 A capacitor is charged by connecting it to a battery and is then disconnected from the battery by opening a switch. When a wire is inserted, connecting the charged plates:

(a) a spark jumps between the plates.

(b) the charge on both plates becomes zero.

(c) the charge on the plates is reversed.

(d) the potential difference between the plates becomes zero.

(e) the kinetic energy of the system is converted to potential energy.

(f) the electrical field strength between the plates becomes zero.

(g) the electrical field strength between the plates is reversed in direction.

(h) None of the above.

CHAPTER 7

Electromagnetism

Note to the instructor: Most students need drill exercises, both in homework and on tests, with the mnemonics concerning the basic electromagnetic phenomena. Few texts provide enough of these exercises to help fix the rules in student memories even within the immediate time interval of an ongoing course. Illustrative exercises, ranging from exceedingly simple to slightly more sophisticated, are given here. Such questions could easily be programmed on computers and could provide most of the needed drill. Variations on the themes illustrated are readily generated.

7.1 The diagram shows the cross section of a wire carrying conventional positive electric current away from us into the plane of the paper.

    • A

(a) By means of an arrow on the diagram, show the direction in which a compass would point if placed at location A and describe the rule you use to help remember this effect.

(b) Show the direction in which the compass would point at two or three other locations of your own choosing.

Note to the instructor: Variations on the preceding question might include (1) giving the direction of the B-field and asking for the direction of conventional current, (2) placing the current-carrying wire in the plane of the paper and asking for the B-field direction in or out of the plane, (3) given a B-field direction in or out of the plane, asking for the direction of current in the wire. The orientation of the wire should be varied in the plane of the paper; it should not be restricted to horizontal or vertical. Similar variations in the geometry of an arrangement can be invoked in many of the questions that follow.

7.2 A uniform B-field is directed toward the right in the plane of the paper, as shown. A wire W, lying perpendicular to the plane of the paper and seen in cross section, carries electric current. The resultant magnetic field at point D is zero.

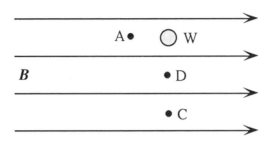

(a) What must be the direction of current in the wire (into or out of the plane of the paper)? Explain your reasoning.

(b) Point A lies the same distance from the center of the wire as point D. Choosing a length for the vector representing the horizontal B-field, construct a vector diagram showing the resultant **B** vector at point A. Explain your reasoning.

(c) Point C lies twice the distance from the center of the wire as point D. Construct a vector diagram showing the resultant **B** vector at point C. Explain your reasoning.

7.3 Two wires lie perpendicular to the plane of the paper and carry equal electric currents in the directions shown. Point P is equidistant from the two wires.

(a) Construct a vector diagram showing the resultant **B** vector at point P. Explain your reasoning.

(b) Suppose a third wire carrying equal current into the plane of the paper were located at P. What would be the direction of the force on this wire? Explain your reasoning.

7.4 Here is a horseshoe magnet with its north and south poles indicated. Between the poles is the cross section of a wire carrying conventional positive electric current toward us out of the plane of the paper.

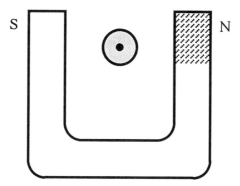

(a) Add to the diagram a few magnetic field lines showing the direction of the B-field (due to the magnet) in the neighborhood of the wire.

(b) Show the direction of the force acting on the wire under these circumstances and describe the rule you use to help remember the experimental facts of the interaction involved.

(c) Does the magnet experience a force under these circumstances? If so, in what direction? If not, why not?

(d) Suppose the wire is rotated so that it lies in the plane of the paper, with the current directed from right to left. What would be the direction of force on the wire?

(e) Suppose you are told that a wire carrying current *into* the plane of the paper experiences an *upward* force. What do infer must be the direction of the B-field to which the wire is subjected?

7.5 The following figure shows a rectangular coil of wire connected to a battery through a switch S. The polarity of the battery is indicated. The coil is located in a uniform B-field directed into the plane of the paper as indicated, and the plane of the coil is in the plane of the paper.

(a) Show the direction of conventional positive current in the coil when switch S is closed, indicating how you arrive your conclusion.

(b) Show by means of arrows the direction of the force acting on the coil at each of the positions A, B, C, and D when the switch S is closed. If the force is zero at any point, say so explicitly. (Show a force directed into the plane of the paper by the symbol ⊗ and a force directed out of the plane of the paper by the symbol ⊙ if needed.)

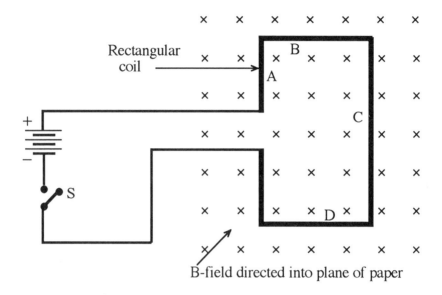

B-field directed into plane of paper

(c) In this orientation relative to the B-field, does the coil tend to rotate or does it tend *not* to rotate? Explain your reasoning. If the coil *does* tend to rotate, indicate the axis and the direction of rotation. If the coil does not tend to rotate in the given orientation, describe the overall effect of the forces to which it is being subjected. If the coil does not tend to rotate in this orientation, describe an orientation in which it *would* tend to rotate.

(d) Answer the questions in part (c) for the case in which the current in the system is reversed.

Note to the instructor: An obvious variation on Problem 7.5 is to ask essentially the same sequence of questions for the case in which the B-field is parallel to the plane of the coil.

7.6 The following diagram shows a solenoid connected to a battery; the axis of the solenoid is parallel to the plane of the paper, and the magnetic field at the end of the solenoid is therefore also parallel to the plane of the paper. Located in the magnetic field of the solenoid is a suspended current-carrying coil, the plane of which is parallel to the magnetic field. The direction of conventional positive current *I* in the coil is shown, as are the directions of the forces (out of the paper and into the paper) on either side of the coil, defining the direction of torque on the coil.

(a) From the information given, deduce the polarity of the battery, explaining your steps of reasoning as you go along. Label the polarity on the diagram.

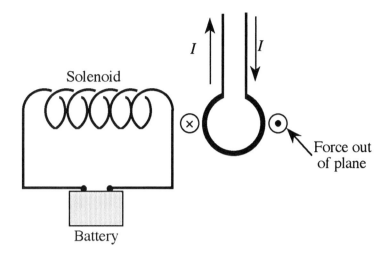

7.7 Two solenoids A and C are sufficiently close together that the magnetic field formed in A, in the presence of electric current, also penetrates into C. (The effect can be greatly strengthened by passing a soft iron rod through both solenoids.)

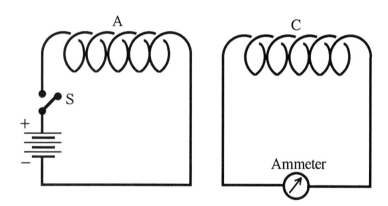

(a) We start with switch S closed so that a steady current is present in solenoid A. Establish the direction of the B-field penetrating into solenoid C and show its direction by means of an arrow labeled ''B-field.'' Explain how you arrive at your result. If the B-field is zero, say so explicitly and explain your reasoning.

(b) While the switch is closed and the steady current is present in A, what is the direction of induced current in C? Show by means of a labeled arrow and explain how you arrived at your conclusion. If the current is zero, say so explicitly and explain your reasoning.

(c) The switch is now opened, and the current in A drops to zero. Describe what, if anything, happens in C, showing the direction of any possibly induced current. Explain how you arrived at your conclusion. What is the situation in C after the current in A has dropped to zero? Explain your reasoning.

7.8 The north pole of a magnet is thrust downward into a horizontally oriented copper ring as follows:

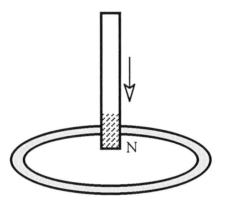

(a) Does the ring experience a force under these circumstances? If so, in what direction? Explain your reasoning.

(b) Does the magnet experience a force? If so, in what direction? Explain your reasoning.

7.9 A solenoid is connected to a battery as shown. Switch S is now closed, and there is a current in the circuit.

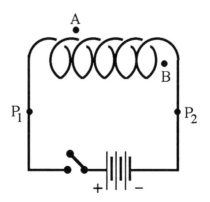

(a) By means of appropriately placed and labeled arrows, show the direction of conventional positive current I in the circuit.

(b) Suppose compasses are placed at points A and B near the solenoid. Show by means of labeled arrows the direction in which a compass would point at each location, and explain how you arrived at your conclusions.

(c) Suppose an unmagnetized iron bar is placed near point B. Will the bar experience a force? If so, in what direction relative to the plane of the paper? If not, why not?

(d) Suppose a resistor is now connected between points P_1 and P_2 with switch S still closed. What will happen to the strength of the B-field at point B? Will it increase, decrease, or remain unchanged? Explain your reasoning.

7.10 The diagram shows a resistive, current-carrying circuit on the right and a closed wire loop on the left. Suppose the contact with the resistor R is moved to the right in the right-hand circuit.

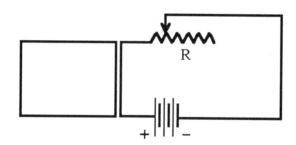

(a) Will current be induced in the wire loop in the left-hand part of the diagram? If so, in what direction? If not, why not? Explain all steps of your reasoning.

7.11 Consider the following diagram. A solenoid produces a B-field directed into the plane of the paper. The current in the solenoid is increasing at a uniform rate, causing the total flux directed into the plane of the paper to be increasing at a uniform rate. Two identical flashlight bulbs, 1 and 2, having resistance denoted by r, are connected in a circuit surrounding the solenoid as shown and light up while the flux increase is taking place. Switches S_1 and S_2 are initially open.

(a) What is the direction (clockwise or counterclockwise) of the emf ε induced in the circuit containing the bulbs? Explain how you arrive at your answer.

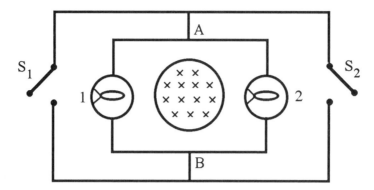

(b) In terms of the quantities ε and r, what is the current in the circuit containing the bulbs while the current in the solenoid is increasing? How do the brightnesses of the bulbs compare under these circumstances? Explain your reasoning.

(c) What happens to the brightness of bulb 2 if bulb 1 is removed from its socket while the current in the solenoid is still increasing? Explain your reasoning.

(d) Suppose that with both bulbs in their sockets, switch S_1 is now closed while S_2 is left open. What happens to the brightness of each bulb while the current in the solenoid is increasing? (Explain your answer in terms of what happens to the combined resistance on the left-hand side of the solenoid.) In terms of the quantities ε and r, what is now the current in bulb 2? The current in bulb 1?

(e) Suppose that switch S_1 is opened and switch S_2 is closed. Answer the same questions asked in part (d).

(f) Note that, in both parts (d) and (e), we are introducing a wire between points A and B and are thus short-circuiting *both* bulbs regardless of which switch we close. In ordinary batteries-and-bulbs circuits previously encountered, a short across any combination of bulbs would have caused *all* the short-circuited bulbs to go out. In the system now under consideration, this is not the case. What happens to the brightness of one of the seemingly short-circuited bulbs is determined by the *direction* taken by the short-circuiting wire. Explain in your own words how this profound difference between the two types of circuit arises. (We sometimes speak of the situation involving the solenoid and the induced emf as constituting a "multiply connected" as opposed to a "simply connected" region.)

7.12 Consider the same situation described in the preceding question except that a third identical bulb (3) is introduced in the manner shown, and the current in the solenoid is *decreasing* at a uniform rate while the B-field is still directed into the plane of the paper.

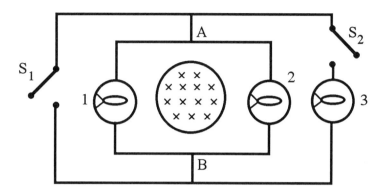

(a) What is the direction of the induced emf ε in the inner circuit? Explain your reasoning.

(b) In terms of the quantities ε and r, what is the current in bulbs 1 and 2 while the current in the solenoid is decreasing and both switches are open?

(c) Suppose that switch S_2 is now closed while S_1 is left open. What will happen to the brightnesses of bulbs 1 and 2 compared to their initial levels? How will the brightnesses of 1, 2, and 3 compare with each other after S_2 is closed? Explain your reasoning.

(d) What is the total resistance of the circuit before S_2 is closed? What is the total resistance after S_2 is closed? What will be the current in bulb 1 after S_2 is closed? (Answer: $2\varepsilon/3r$) What will be the current in bulb 2 after S_2 is closed? (Answer: $\varepsilon/3r$) Explain your reasoning. Are your results consistent with your qualitative predictions regarding comparative brightnesses in part (c)?

(e) Answering the same questions as in parts (c) and (d), discuss what will happen if switch S_1 is closed while S_2 is left open.

(f) What will happen if both switches are closed? How will current and brightnesses compare with the situation that obtains when both switches are open? Explain your reasoning.

7.13 A long solenoid lies with its axis perpendicular to the plane of the paper. A linearly changing current in the solenoid induces a clockwise emf. A wire of negligible resistance connects three identical flashlight bulbs in the manner shown. The bulbs light when the emf is induced.

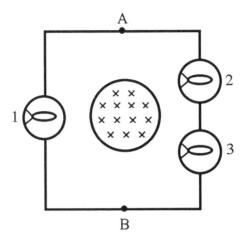

(a) How do the brightnesses of the bulbs compare with each other under the circumstances illustrated in the diagram? Explain your reasoning.

(b) What will happen to the brightness of each bulb if bulb 2 is removed from its socket? Explain your reasoning.

(c) Restore bulb 2 to its socket. What will happen to the brightness of each bulb if a wire shorts points A and B while connected so that it lies around the right-hand side of the solenoid? How will the brightnesses now compare with those that obtained before the shorting? Explain your reasoning.

(d) Answer the same question as part (c) if the wire shorts points A and B while lying around the left-hand side of the solenoid. Explain your reasoning.

(e) Answer the same question as part (c) if the wire shorts points A and B by being connected symmetrically through the center of the solenoid. Explain your reasoning.

(f) Suppose that two voltmeters are connected between points A and B. One voltmeter (call it V_1) is connected so that it and its wires make a circuit around to the left of bulb 1, while the other meter (call it V_{23}) is connected so that it and its wires make a circuit around to the right of bulbs 2 and 3. Note that both voltmeters are connected between the *same* two points A and B. In ordinary batteries-and-bulbs circuits, we found that voltmeters connected between the same two points show the same readings no matter what the geometry of the arrangement. Will voltmeters V_1 and V_{23} in this arrangement show the same readings? Why or why not? Explain, as though you were helping a fellow student see what is involved, why it is necessary to be very careful about interpreting voltmeter readings in circumstances that involve induced emf's and multiply connected regions. [If you are interested in pursuing further the issue of the meaning of voltmeter readings, read the article by R. H. Romer "What Do 'Voltmeters' Measure? Faraday's Law in a Multiply Connected Region," *American Journal of Physics*, **50**, 1089 (1982).]

7.14 The following circuit consists of four identical flashlight bulbs, each with resistance r, and an ideal battery (negligible internal resistance relative to the bulbs) that maintains a constant potential difference ΔV across its terminals. A solenoid oriented perpendicular to the plane of the paper and producing a B-field directed into the plane of the paper occupies the position shown.

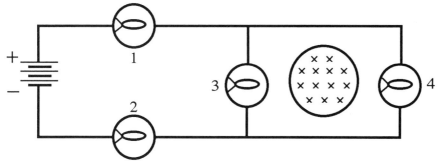

(a) Initially, with the current in the solenoid constant and the B-field unchanging, find the current in each of the four bulbs in terms of ΔV and r. Explain your solution step by step. How do the brightnesses of the bulbs compare with each other?

(b) Suppose we now either increase or decrease the current in the solenoid linearly with time. What direction of induced emf ε (clockwise or counterclockwise) do we need to produce around the solenoid to make bulb 3 go out (i.e., to carry zero current)? Explain your reasoning.

(c) What emf ε (expressed in terms of the known potential difference ΔV) must be induced to produce zero net current in bulb 3? Show and explain all steps of your reasoning. [Hint: Calculate the effective resistance lying to the left of the solenoid and the total resistance in the loop around the solenoid; then find the induced current in the loop and the induced current through bulb 3.] (Answer: $\varepsilon = \Delta V/2$)

(d) What is the current in each of the bulbs 1, 2, and 4 under the circumstances obtaining in part (c)? Show and explain all steps of your reasoning. (Answer: $\Delta V / 2r$)

(e) For the situation in part (c), what will be the current in each remaining bulb if bulb 4 is removed from its socket? Explain your reasoning.

(f) For the situation in part (c), what will be the current in each bulb if a wire shorts bulb 4 by being connected around the right-hand side of the solenoid? Show your reasoning. Answer the same question for the case in which a wire shorts bulb 3 by being connected around the left-hand side of the solenoid.

(g) What magnitude and direction of emf would cause bulb 4 to go out rather than bulb 3? Show and explain all steps of your reasoning.

Note to the student: In the following questions, circle the letter or letters designating statements you believe to be <u>correct</u>. <u>Any number</u> of statements may be correct, <u>not</u> necessarily just one. You must examine each statement on its merits.

7.15 If the north pole of a magnet is thrust downward into a horizontally oriented copper ring as shown in the following diagram, the ring will experience

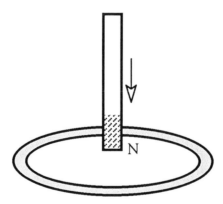

(a) a downward force.

(b) an upward force.

(c) a sidewise force.

(d) zero force.

(e) a clockwise torque as seen from above.

(f) a counterclockwise torque as seen from above.

(g) None of the above.

7.16 The diagram shows a ring of copper with its plane perpendicular to the horizontally oriented axis of the nearby cylindrical magnet. A current will be induced in the ring if

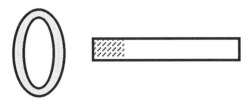

(a) the magnet is moved toward the ring.

(b) the ring is moved away from the magnet.

(c) the ring is rotated about any of its diameters.

(d) the magnet is spun around its horizontal axis (the line through its length.)

(e) the magnet is rotated around a vertical axis through its center.

(f) the magnet is rotated around a horizontal axis perpendicular to the plane of the paper.

(g) the magnet is moved up or down.

(h) the ring is rotated around its center in the plane in which it lies.

(i) None of the above.

7.17 A small bar magnet lies in a region of nonuniform magnetic field. Under such circumstances, depending on how it is oriented relative to the field direction, the magnet might experience

(a) a torque but zero net force.

(b) a net force but zero torque.

(c) both a torque and a net force.

(d) neither a torque nor a net force.

(e) translational motion if it is free to move.

7.18 In the figure, we are looking down on a bare wire ABCD lying on a table top. Lying across ABCD and making good electrical contact with it is another bare wire EF which is free to move. A B-field is directed down through the rectangle of wires. If the strength of the B-field is decreased, the wire EF will

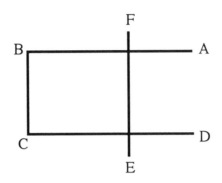

(a) remain stationary.

(b) slide to the right.

(c) slide to the left.

(d) rotate clockwise.

(e) tend to move upward out of the plane of the paper, breaking contact with ABCD.

(f) tend to move downward into the plane of the paper, pressing against ABCD.

(g) None of the above.

7.19 The helical coil of wire carries an electric current. As a result of the current, the coil

(a) tends to unwind.

(b) tends to become shorter.

(c) tends to become longer.

(d) has no tendency to alter its shape.

(e) behaves in a way that cannot be predicted since effect depends on direction of current.

CHAPTER 8

Particle Trajectories in E- and B-Fields

8.1 In an evacuated space, an electron moves with velocity v toward the left in the plane defined by the paper as shown. The electron enters a region of uniform B-field directed out of the plane of the paper. The magnetic field strength B drops very sharply to zero outside the region defined by the dashed lines. Note that, depending on the strength of the field, the electron might exit into the field-free region at various points; be sure to take this into account when sketching trajectories.

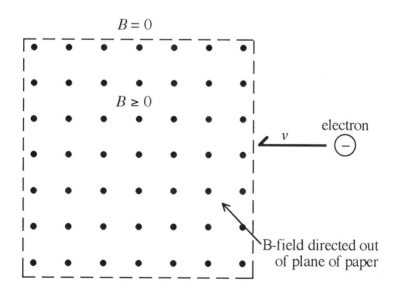

(a) Sketch possible trajectories of the electron for each of the following four B-field conditions within the region defined by the dashed lines and label each trajectory with the corresponding number:

(1) $B = 0$ throughout the region.   (3) Moderately strong B-field.

(2) Weak, but not zero, B-field.   (4) Very strong B-field.

8.2 A uniform B-field is directed into the plane of the paper in an evacuated space. A hydrogen ion (H^+) and an alpha particle (He^{2+}) both have the same velocity v (in the plane defined by the paper) at point P. An alpha particle has four times the mass of a hydrogen ion. The field is so strong that in the absence of collisions, both particles execute complete circular trajectories within the region of the field.

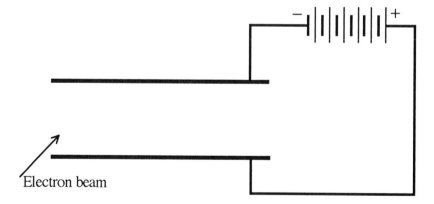

(a) Deduce the relative size of the circles and sketch them to scale on the diagram, showing the direction of travel in each case. Explain your reasoning.

8.3 A beam of electrons in a highly evacuated space enters an E-field between capacitor plates in the direction shown in the diagram.

(a) Sketch a possible electron trajectory for each of the following cases of different E-field strength, remembering that gravitational effects are unobservably small. Sketch each trajectory neatly and carefully—not sloppily and carelessly—and label each one with the appropriate number. (If the beam exits from between the plates, be sure to show the character of the trajectory after the exit.)

(1) E = 0, i.e., plates uncharged.

(2) Weak E-field, small potential difference across the plates.

(3) Moderate E-field.

(4) Very strong E-field.

8.4 In part (a) of the diagram, a negatively charged particle moving at velocity v in an evacuated space enters a region between capacitor plates as shown. In part (b), a positively charged particle enters a sharply defined region of uniform B-field directed into the plane of the paper.

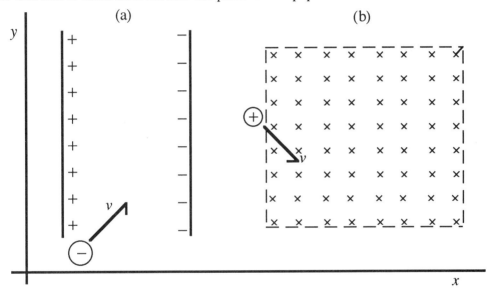

(a) Examine the following statements. Indicate whether each is true or false and explain the reasoning behind your conclusions.

   (1) The negatively charged particle in (a), on entering the electrical field between the capacitor plates, moves without change in the y-component of its velocity.

   (2) The negatively charged particle in (a), on entering the electrical field between the capacitor plates, moves without change in the x-component of its velocity.

   (3) The positively charged particle in (b), on entering the region of magnetic field, moves without change in the x-component of its velocity.

(4) The positively charged particle in (b), on entering the region of magnetic field, moves without change in the y-component of its velocity.

(5) Neither particle changes its speed as it continues along its trajectory.

8.5 In the following diagram, an electron, at the position indicated, moves with velocity v in the direction shown. The space is highly evacuated, and a uniform magnetic field of strength B is directed out of the plane of the paper. B drops sharply to zero immediately outside the region shown. Note that the electron may exit the region of the B-field; be sure to take this possibility into account when sketching trajectories.

(a) Sketch a possible trajectory of the electron, showing the direction of travel. Explain how we know that the trajectory within the B-field is circular.

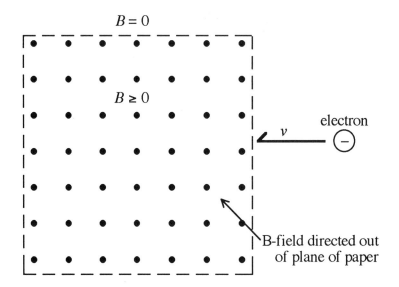

(b) Explaining all steps, derive an expression for the radius R of the circular part of the trajectory in terms of known quantities such as the mass and charge of the electron, the velocity v, and the magnitude of the field strength B.

(c) Interpret the expression you have obtained for R by adding to the figure:

(1) A possible trajectory of the electron with a velocity v very much *smaller* than that supposed in your initial trajectory (all other parameters held fixed). Label this trajectory v_{small} and explain your reasoning.

(2) A possible trajectory that might be observed with a very much weaker B-field (all other parameters unchanged). Label this third trajectory B_{weak}. Explain your reasoning.

(d) Under the circumstances sketched, is it possible for the particle to make a complete circle within the B-field? Why or why not?

8.6 A positively charged particle moves in the positive x direction in a uniform magnetic field directed into the plane of the paper as sketched in the following diagram. The resultant force on the particle can be made zero by introducing a uniform electric field of appropriate strength in the

(a) $+y$ direction.

(b) $-y$ direction.

(c) $+x$ direction.

(d) $-x$ direction.

(e) $+z$ direction (out of plane of paper).

(f) $-z$ direction (into plane of paper).

(g) None of the above; an electric field cannot cancel a magnetic field.

CHAPTER 9

Wave Phenomena

9.1 A straight wave train in a ripple tank is incident at an interface between shallower and deeper water.

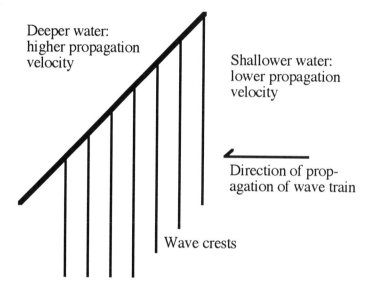

(a) Add to the diagram (1) an incident ray, (2) a reflected ray, and (3) a refracted ray, labeling each one with the corresponding number.

(b) Redraw the diagram and add (4) two or three reflected wave fronts and (5) two or three refracted wave fronts, labeling each set with the corresponding number.

(c) If the frequency of the incident wave train is ν_i, how will the frequency ν_r of the reflected wave train compare with ν_i? Will it be greater than, less than, or equal to ν_i? How will the wavelength λ_r

of the reflected wave compare with the wavelength λ_i of the incident wave? Explain your reasoning.

(d) How will the frequency ν_t of the transmitted (or refracted) wave train compare with ν_i? How will the wavelength λ_t of the transmitted wave compare with λ_i? Explain your reasoning.

9.2 Is the following statement true or false? Explain your reasoning. "In a two-source interference pattern in a ripple tank, the nodes and antinodes all move out to larger angles from the principal axis as the *frequency* of oscillation of the sources is decreased without change in separation of the sources."

9.3 In this rough sketch of a two-source interference pattern in a ripple tank, the two sources S_1 and S_2 are oscillating in phase with each other. The open arcs represent crests; the shaded arcs represent troughs; the regions of constructive and destructive interference are indicated.

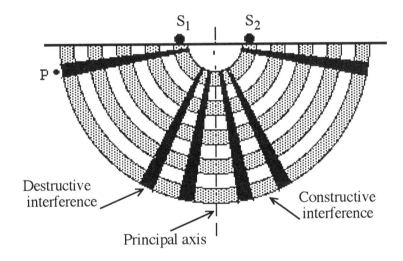

(a) Is the sketch internally consistent or inconsistent? Explain your reasoning. (Hint: Look at the comparative magnitudes of the spacing $S_1 S_2$ between the sources and the wavelength of the ripples. Does the diagram make qualitative sense for the magnitudes indicated? Why or why not?)

(b) Consider the position denoted by the dot at P. What is the difference, at this point of observation, between the lengths PS_2 and PS_1 (i.e., the length $PS_2 - PS_1$) expressed in wavelengths λ of the ripples? Explain your reasoning.

(c) Sketch what this pattern might look like if the spacing S_1S_2 between the sources were decreased somewhat without any change in the wavelength of the ripples or the phasing of the sources.

(d) Sketch what this pattern might look like if the wavelength λ of the ripples were increased somewhat without change of spacing between the sources or in their phasing.

(e) Sketch what this pattern would look like if the source spacing and ripple wavelength were unchanged but the sources were oscillating exactly out of phase with each other.

9.4 In the diagram we are looking down on a two-source interference pattern in a ripple tank; the two point sources oscillate in unison. (The source positions S_1 and S_2, the principal axis, and a position P are shown. The ripples themselves are not shown.) It is given that point P is located on the second line of destructive interference (antinode) away from the principal axis.

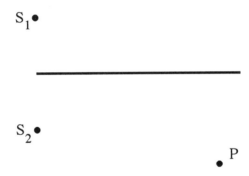

(a) Use a ruler to measure the relevant distances directly on the diagram, and, from these distances, calculate the wavelength of the ripples making the pattern. Explain your reasoning.

9.5 The following sketch represents a ''photograph'' of a two-source interference pattern in a ripple tank, a photograph that is *reduced* in scale relative to the tank pattern itself. The two sources S_1 and S_2 are known to be oscillating exactly *out* of phase with each other. The actual wavelength in the tank is known to be 2.0 cm. The observation point P is known to be on the second locus of *destructive* interference away from the principal axis; i.e., there is one other locus of destructive interference between this one and the principal axis.

(a) Using a ruler on the ''photograph'' and from information given above, calculate the *scale* of the photograph. How many centimeters in the photograph correspond to 1 cm in the ripple tank? Explain your reasoning.

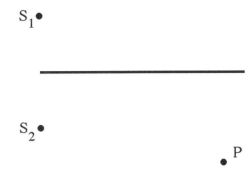

(b) Calculate the actual separation between the sources in the tank. Explain your reasoning.

9.6 In describing sinusoidal wave trains, we always encounter expressions such as $\sin(2\pi x/\lambda)$ or $\cos(2\pi x/\lambda)$,where x denotes position along a coordinate axis and has dimensions of length. Explain in your own words where the $2\pi/\lambda$ comes from. Why not simply $\sin x$ or $\cos x$?

9.7 A horizontal, stretched string carries a sinusoidal wave train described by the following equation:

$$y = A\sin(ax + bt)$$

(a) In what direction along the x-axis is the wave moving? Explain your reasoning.

(b) What is the physical meaning of A? Explain how you arrive at this interpretation.

(c) In terms of a and/or b, what is the wavelength of the wave train? Explain your reasoning.

(d) In terms of a and/or b, what is the frequency of the wave train? Explain your reasoning.

(e) In terms of a and/or b, what is the time taken for a particle of the string to go through one complete cycle of its up-and-down motion? Explain your reasoning.

(f) In terms of a and/or b, what is the propagation velocity of the wave train? Explain your reasoning.

(g) In terms of a and/or b, at what values of x are the deflections of the string zero at the instant $t = 0$?

(h) What would be the effect of changing the plus sign in $(ax + bt)$ to a minus sign? Explain how you arrive at your answer.

9.8 A pulse of the shape shown propagates to the right along a string with velocity V. Describe *in detail* the motion of particle A on the string as the pulse goes by. Be sure to indicate (1) the direction in which A is moving at various points along the pulse and (2) whether the velocity of A is increasing, decreasing, or zero at these various points. (Points 1 through 5 on the diagram are suggested as convenient references for the discussion.)

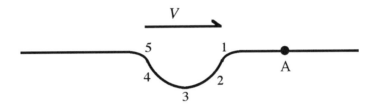

9.9 A *rarefaction* pulse (no compression phase) travels from right to left along this coil spring (or "slinky").

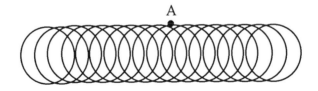

(a) Describe what is happening to the coils as the pulse goes by, and describe the motion of point A on one of the coils, starting just as the leading edge of the pulse arrives at A until the time the trailing edge passes. Include in your description the direction of motion of A, any changes in direction of motion, and the direction, as well as any changes in magnitude, of the velocity of point A. Does point A return to its initial position after passage of the pulse or is it permanently displaced? If A is permenently displaced, what altered shape of pulse would cause it to return to its initial position? Explain your reasoning as you go along.

(b) What differences do you see between the particle motion you described in part (a) and the particle motion undergone by point A on the string in the case of passage of a negative transverse pulse such as the following?

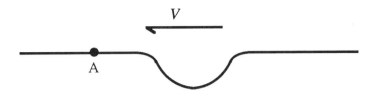

9.10 Consider the following case in which two transverse pulses of identical shape but opposite phase travel in opposite directions along a string and interfere destructively.

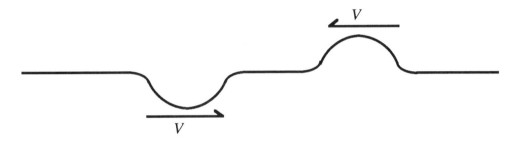

(a) At the instant the two pulses are exactly superposed, the deflection of the string is *everywhere* zero. Remember that each pulse is propagating energy and momentum in the direction in which it is traveling. What has happened to the momentum and energy at the instant of superposition? Note that their disappearance would be a violation of the conservation laws. Think carefully about the *particle* velocity at each point along the string at the instant of zero deflection, and indicate its direction at various points. Explain your reasoning.

(b) Is the following statement true or false? "At the instant the two pulses cancel each other's deflections destructively, *all* the energy of the two waves is present in the form of kinetic energy of particle motion of the string." If the statement is false, alter it into a true statement; if you consider it true, explain why.

(c) Perform an analysis parallel to that in (a) and (b) for the instant at which two identical *positive* pulses, propagating in opposite directions, superpose exactly and interfere constructively. In what form, at this instant, does one find the energy that is being propagated? Explain your reasoning.

9.11 For the sake of simplifying a useful exercise, let us imagine the following sharply stepped pulses A and B, propagating in opposite directions along a stretched string. (It is, of course, not physically possible to generate such idealized, perfectly rectangular pulses.) We imagine two photographs of the pulses: one taken at instant t_1, just as the pulses are about to overlap (interfere), and the second taken at instant t_2, just as they are about to separate after interfering.

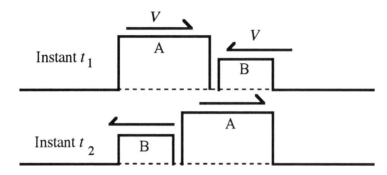

(a) Take three instants of time, approximately equally spaced between instants t_1 and t_2, and sketch the shape that the string would assume at each of those instants as pulses A and B overlap.

9.12 Transverse pulses of the shapes shown here propagate along strings in the directions shown. The end of each string is fastened to a rigid wall. Sketch the shape of the reflected pulse that would "emerge" from the imaginary (dashed) continuation of the string on the other side of the wall in each case. Explain the reasoning behind your sketches.

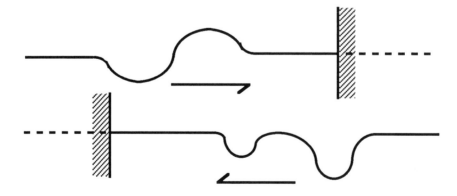

Note to the instructor: Problems exactly like 9.12 should be posed with respect to reflections of transverse waves at perfectly free boundaries and at boundaries between strings of different mass per unit length. Similar problems should be posed with respect to reflections of longitudinal wave pulses.

9.13 In a ripple tank, it is easy to make a straight wave pulse by quickly moving a rod either backward or forward near the water surface. Consider a ripple tank in which such a low amplitude pulse, in the form of a crest, is propagating from left to right and undergoes normal incidence at the end of the tank. (In these figures we are looking, from the side, at a cross section through the tank. Note that the tank has a fairly wide rim.) In case 1, the tank is only partly filled, and the ripple is incident at a vertical wall. In case 2, the tank is completely filled, and the ripple is incident at the rim.

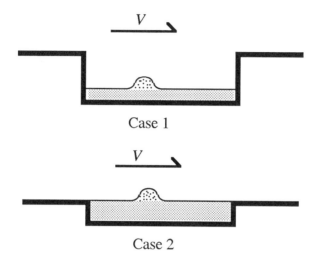

Case 1

Case 2

(a) In the light of what you have learned about reflection of waves at various boundaries, predict the shapes of the reflections that would be observed in the two cases illustrated. Would the reflections be identical or would they be different? Explain your reasoning in making your predictions.

(b) Noting that the ripple is a transverse wave, compare the reflections predicted in the two cases above with (1) the reflections, at fixed and free boundaries, of transverse pulses on a string and (2) the reflections of longitudinal pulses, at fixed and free boundaries on a spring (or "slinky"). To which of the phenomena, 1 or 2, are the ripple reflections most nearly analogous?

9.14 Given a stretched string, describe how you would generate (a) a transverse pulse with only positive deflection (i.e., only a positive phase); (b) a transverse pulse with only negative deflection (i.e., only a negative phase); (c) a transverse pulse with both positive and negative deflections (i.e., with both positive and negative phases). In your description, indicate what motion you would actually execute with your hand at one end of the string.

9.15 Given a stretched coil spring or "slinky" (say, lying on the table), describe how you might generate (1) a longitudinal pulse with only a positive phase (compression); (2) a longitudinal pulse with only a negative phase (rarefaction); and (3) a longitudinal pulse with both positive and negative phases. In your description, indicate what motion and displacement you would actually execute with your hand at one end of the slinky.

    (a) Now give a parallel description regarding generation of transverse pulses on a string: What motion and displacement do you execute with your hand at one end of the string to generate either a purely positive or a purely negative pulse?

    (b) Identify the very significant differences between the displacements you execute for transverse waves on the one hand and longitudinal waves on the other when it comes to generating pulses having only one phase, positive or negative.

9.16 Using appropriate sketches, describe what is meant by a "continuous wave train" in contrast to a single pulse. Describe what is meant by a "periodic" wave train. Describe what is meant by a "sinusoidal" wave train. What is the difference in meaning between the terms "periodic" and "sinusoidal"?

9.17 With the help of appropriate diagrams, define the terms "frequency, f " and "wavelength, λ " of a periodic wave train. Then, reasoning *arithmetically* from the definitions, establish the relationship among the three quantities V (velocity of propagation), f, and λ. (In other words, be able to reason out the relationship whenever you need it rather than memorizing it as a rigid formula.)

9.18 Define the concepts "ray" and "wave front," using diagrams as well as words, and illustrating the concepts in the cases of both straight and circular pulses.

9.19 Using appropriate diagrams, define the concepts "angle of incidence" and "angle of reflection" for both straight and circular pulses incident at a straight rigid barrier. Define what is meant by "normal" and "glancing" incidence. Sketch and label the angles in wave front as well as ray representations. Sketch, in both ray and wave front representations, what happens as the angle of incidence is increased from normal to glancing.

9.20 Given a situation in which a straight wave pulse propagating in a region of deeper water (higher propagation velocity) is incident at a straight interface with a region of shallower water (lower propagation velocity). Sketch separate ray and wave front diagrams showing the incident, transmitted, and reflected pulses. Sketch such diagrams for the case in which the situation is reversed and the incident wave pulse arrives in the shallower region. In connection with your diagrams, define the concept "angle of refraction" (or "angle of transmission").

9.21 In each of the instances and diagrams arising in the preceding question, sketch how the angle of refraction changes as the angle of incidence is varied from normal to glancing.

9.22 On the basis of your observations with circular wave pulses, sketch what happens to the wave front in reflections from a straight rigid barrier, i.e., show how the reflected wave changes as you move the center of the incident wave closer to, or farther from, the barrier.

9.23 Sketch what happens to wave fronts when a circular pulse is incident at a straight *refracting* interface, making diagrams that show different distances of the center of the incident circular pulse from the interface.

9.24 Suppose a straight sinusoidal wave train arrives at normal incidence to a straight barrier that is shorter than the length of the wave fronts (i.e., the unimpeded portion of the wave front can propagate past the barrier while part of the wave front is blocked). Sketch the pattern to be observed in the region beyond the barrier for the cases in which the wavelength λ of the wave train is (1) very short relative to the length of the barrier, (2) very long relative to the length of the barrier, and (3) of intermediate length.

9.25 A straight sinusoidal wave train arrives at normal incidence to a straight barrier that contains an opening of width D, and the waves are blocked except for passage through the opening. Sketch the pattern of wave fronts transmitted through the opening for different ranges of the ratio of wavelength to opening width λ/D (i.e., for small, large and intermediate values of λ/D).

9.26 With the help of appropriate sketches, define the concepts "refraction," "diffraction," and "interference" and explain how they differ.

9.27 You have probably seen a bow wave on the water surface generated by a rapidly moving boat or ship. Let us do some qualitative thinking about the circumstances in which bow waves are formed.

(a) When a boat moves through the water, waves are invariably generated on the water surface. Suppose the boat moves very, very slowly. How does the velocity of the waves compare with the velocity of the boat under these circumstances? How do the waves behave relative to the boat? Does a bow wave form?

(b) Suppose the boat now accelerates, slowly increasing its speed. What happens to the spacing between the leading waves and the bow of the boat? What speed must the boat attain so that the waves "stick" to the bow? What, in general, are the circumstances under which bow waves are formed? How does the shape of the waves formed by the slowly moving boat compare with the shape of the bow wave?

(c) If the opportunity arises, reproduce the situation we have just been discussing by moving a vertically oriented pencil or rod through a ripple tank in the laboratory or a basin of water at home, and observe what happens at slow and at increasingly rapid motions. What is the shape of the ripples formed when the rod moves slowly? What is the shape of the bow wave? How do you account for the shape of the bow wave?

(d) In the light of the preceding thinking and reasoning, try to visualize the effect of moving objects on the air that surrounds them. How would the compression in front of a moving object behave relative to the object? Under what circumstances might a "bow wave" form? How would the effect of a baseball compare with that of a bullet?

9.28 The shaded line represents a boundary between a deeper and a shallower layer of water in a ripple tank. (We are looking down on the surface of the water.) The velocity of propagation of ripples is higher in the deeper water. Suppose you were to take a pencil and move its point rapidly (faster than the propagation velocity of ripples on either side) in the water right along the boundary between the two depths, starting at the left-hand side.

Deeper

Shallower

(a) Sketch the bow wave pattern that would be observed in each region (on either side of the depth change) at the instant the pencil point is at the right-hand end of the diagram. Explain your reasoning.

9.29 The diagram represents the arrival of a straight wave front at an interface between two regions of differing propagation velocity. The diagram also shows the refracted wave front.

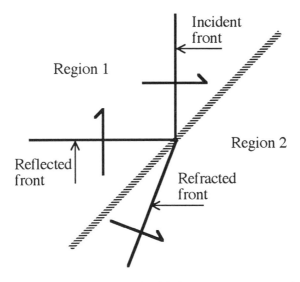

(a) Which region, 1 or 2, has the higher wave propagation velocity? Explain your reasoning.

9.30 A straight wave train is propagating at velocity V_1 from left to right, in a ripple tank. The parallel solid lines represent crests of the train of ripples. The train is normally incident at a step in depth, the depth being shallower to the right and the propagation velocity V_2 lower. Let us suppose that V_2 is about half of V_1 (such a ratio would be hard to attain physically).

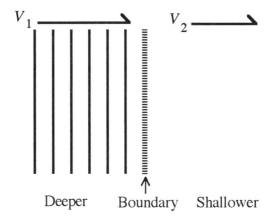

(a) Sketch the spacing between the crests that would be observed on the right-hand side of the step change in depth and propagation velocity. Explain your reasoning.

(b) Explain why it is that the *frequency* of the wave train, rather than the wavelength, changes under such circumstances. (Keep this fact in mind for future reference in the more subtle case of change in velocity of propagation of light.)

9.31 A straight wave train is propagating at velocity V from left to right in a ripple tank. The parallel solid lines represent crests of the train of ripples. The wave train is normally incident on a barrier that contains regularly spaced openings. The width of these openings is somewhat less than the wavelength of the wave train. The spacing between the centers of the openings is about twice the wavelength of the wave train. Diffracted wavelets, transmitted through the openings, superpose on the right-hand side of the barrier while the rest of the wave train is blocked.

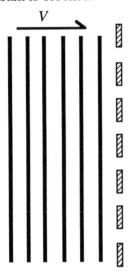

(a) Sketch the pattern of overlapping wave trains you would expect to see on the right-hand side of the barrier under these circumstances as far as the first-order interference. Note that the pattern is *very* different from that of the two-source interference pattern. The superposition of circular wavelets leads to *straight* wave fronts propagating into the right-hand region. You can actually see the pattern of the straight wave fronts if you set the situation up carefully in a ripple tank. Make the wavelength as long as feasible and design the barrier accordingly. When you watch the pattern, let your eyes sweep from right to left with the waves. You will then see the interference pattern that arises.

9.32 Shock cord is a very stretchable rubber "rope" that is used for highly flexible fastenings (a given segment can be stretched to more than twice its relaxed length). Suppose you start with a length of shock cord fastened at one end and pull on the other end, stretching the cord slightly under relatively low tension. You then make a transverse wave pulse by deflecting the end you are holding and note its propagation velocity V_1. You then stretch the cord by quite a large amount, increasing its length (and the tension) significantly. You now observe a larger propagation velocity V_2.

(a) Explain, in terms of what you have learned about the velocity of transverse waves on a string, why V_2 is greater than V_1. Make your explanation *physical* by considering the application of Newton's second law to a small chunk of the string displaced from its equilibrium position, and not by simply referring to symbols in a formula for velocity of propagation. Note that *two* effects are involved (mass of the chunk and force applied), not just one. Be sure to account for each effect in your explanation.

9.33 A flexible coil spring ("slinky") hangs from the ceiling. You produce a *transverse* pulse by deflecting the lower end. As the pulse propagates up to the ceiling, its propagation velocity changes continuously.

(a) Predict, in terms of what you have learned about the velocity of transverse waves, how the propagation velocity changes on the way up. Does it increase or decrease? Explain why the velocity changes rather remaining the same, as it does in the case of a horizontal string. (Note that *two* effects are involved, not just one. Be sure to account for each effect in your explanation.)

9.34 The following questions extend to phenomena related to sound what you have learned about longitudinal waves by observing their behavior on a coil spring.

(a) In the light of what you have seen of the generation and behavior of compression and rarefaction pulses on a slinky, sketch what you visualize might be happening to the air in a tube when a piston or diaphragm moves rapidly back and forth at one end.

(b) Is it possible to make a sound pulse having only a compression phase and no rarefaction phase by moving the piston inward and then returning it to its initial position? Why or why not? How does this situation compare with the generation of pulses on the slinky?

(c) What variables (ordinate and abscissa) might you use to make a graph representing a sound pulse? There are several possibilities.

(d) How would you imagine the interference of sound waves to take place? Suppose you had two audible point sources of sound, analogous to the situation with two point sources in the ripple tank (e.g., two tuning forks sounding in unison). How would you go about finding regions of constructive and destructive interference using only your own ears as detecting devices? That is, what positions would you explore and what would you listen for? Use a diagram to illustrate your answer.

(e) Predict how the compression and rarefaction phases of sound pulses would be reflected from a rigid wall and explain your reasoning.

9.35 ''Intensity (I)'' of a wave train is defined as the energy transported (in the direction of propagation of the wave train) through a unit area per unit time. Thus the SI units would be watts per square meter per second. Textbooks show that intensity I is directly proportional to the *square* of the amplitude A of the train.

(a) How does the surface area of a sphere vary with the radius R of the sphere? In the light of this functional relation and the constraint of conservation of energy, how must the intensity of a spherically spreading wave train (e.g., a sound wave in air or water) vary with R if the wave spreads without dissipating any of its energy? Explain your reasoning. Qualitatively, how would you expect the intensity to vary with R , in comparison with the variation in the nondissipative situation, if there were small, continuous dissipation taking place (as is, of course, actually the case)?

(b) In the light of the fact that intensity is proportional to the *square* of the amplitude, how would you expect the *amplitude* of the wave train to vary with R in the ideal (nondissipative) situation in part (a)? Explain your reasoning.

(c) Consider a train of water waves on the surface of a pond, generated, for example, by moving a stick up and down at the source. In this case, the wave train is spreading *cylindrically* rather than spherically as in parts (a) and (b). How does the area of a cylindrical surface vary with the cylindrical radius R? In the light of this functional relation and the constraint of conservation of energy, how must the intensity of the cylindrically spreading wave train vary with R in the ideal (nondissipative) case? How must the *amplitude* of the cylindrically spreading wave train vary with R? Explain your reasoning.

(d) Making up some examples of your own with numerical ratios of radial distances, compare the decay of amplitude with distance in the spherical wave with the decay of amplitude with distance in the cylindrical wave, and illustrate the enormous quantitative difference between the effects of spreading in the two situations. What is the origin of the difference in the rate of decay with distance in the two different geometries? Explain your reasoning.

9.36 Use the following equation for the nth harmonic (nth eigenfunction) of a standing wave on a string with fixed ends. (Our clock starts at $t = 0$.)

$$y_n = 2A \sin\frac{n\pi x}{\lambda_n}\cos\omega_n t$$

(a) At what subsequent clock readings t will the string be completely flat (i.e., straight and undeformed)? Present your analysis in clear, sequential mathematical steps with explanations of reasoning as you go along. Express your final result in terms of the *period* T_n of the oscillation. (Note: If you encounter a new sequence of integer numbers, you will need a symbol other than n, since n has been preempted for the designation of eigenstates.)

9.37 An organ pipe has its closed end at the origin ($x = 0$) and its mouth at $x = L$ as sketched.

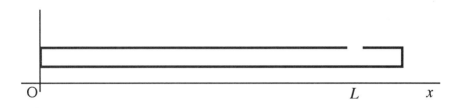

(a) It is asserted that in a specific case of resonance in the pipe, the wave number k has the value 2.28 ft $^{-1}$ and the angular frequency ω has the value 2510 s $^{-1}$. Is this combination of numerical values physically reasonable or unreasonable for a sound wave in air at room temperature? Explain your reasoning.

(b) If the wave described in part (a) is the *second* harmonic of the pipe, what must be the length L of the pipe? Find the numerical value and explain your reasoning.

(c) Using the departure from atmospheric pressure (Δp) as the dependent variable, sketch the standing wave form in (b) on the following set of coordinates (according to the usual convention for depicting standing waves). Indicate the locations of pressure nodes and antinodes.

(d) Which of the following mathematical expressions would most conveniently describe the *standing wave* pressure variation in the pipe with respect to x and t as independent variables, given the coordinate system in part (c)? Circle your choice and explain the reasoning behind it. If more than one of the equations would be equally convenient, say so and explain why.

$$\Delta p = \Delta P_m \cos\left[\frac{2\pi x}{\lambda} \pm \omega t\right]$$

$$\Delta p = \Delta P_m \sin\frac{2\pi x}{\lambda} \sin \omega t$$

$$\Delta p = \Delta P_m \cos\frac{2\pi x}{\lambda} \sin \omega t$$

$$\Delta p = \Delta P_m \sin\frac{2\pi x}{\lambda} \cos \omega t$$

9.38 Following are some functions of the form $f(x \pm Vt)$. Examine each function and assess whether it might be useful for describing a physically possible propagating wave form. Explain your reasoning.

$$y = A(x - Vt)$$
$$y = A(x + Vt)^2$$
$$y = A\sqrt{x - Vt}$$
$$y = A\ln(x + Vt)$$

$$y = 0 \qquad\qquad \text{for } x < 0$$
$$y = A\varepsilon^{-(x-Vt)} \quad \text{for } x \geq 0$$

9.39 Describe *qualitatively* what actually happens to the energy being transported by a transverse wave on a string when the wave is incident at a *real* wall. Is *all* the energy reflected? Why or why not? What happens in the wall? The phenomena taking place in the wall are actually *very* complex, but try to visualize at least some of what might happen.

9.40 Can you generate, on a stretched string, a wave that would transport *angular* momentum as well as linear momentum? Why or why not? If you can generate such a wave, describe how you would do so.

CHAPTER 10

Images with Mirrors and Lenses

10.1 In the following diagram, M is a plane mirror; B is a very small, bright light bulb that can be treated as a point source of light; H is an opaque housing that does not transmit light; and O is a line anywhere along which an observer can stand to try to see the image of the light bulb in the mirror.

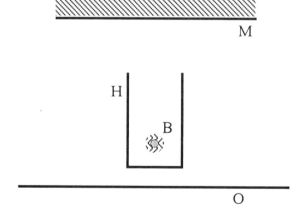

(a) By using relevant rays of light, determine the locations along line O from which the image of B is visible in the mirror and the locations from which it is not visible. Mark these regions accordingly along line O, and explain the reasoning you used in drawing the rays.

(b) Explore how moving B around within the housing H would affect the regions you have mapped out.

10.2 In the following diagram, M is a plane mirror; B is a very small, bright light bulb that can be treated as a point source of light; P is an opaque plate that does not transmit light; and O is a line anywhere along which an observer can stand to try to see the image of the light bulb in the mirror.

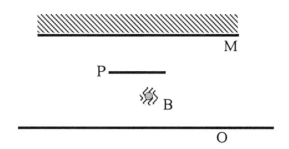

(a) By using relevant rays of light, determine the locations along line O from which the image of B is visible in the mirror and the locations from which it is not visible. Mark these regions accordingly along line O, and explain your reasoning.

(b) Explore how moving B around in the region behind plate P would affect the regions you have mapped out.

Note to the instructor: Many of the following problems on images with lenses would have counterparts with concave and convex mirrors, probing similar levels of understanding. The problems could simply be rephrased accordingly.

10.3 In the two diagrams, give definitions (separately) of each of the principal focal points F_1 and F_2 of a thin *diverging* lens by (a) drawing rays that define the points and (b) describing in your own words how the rays are drawn.

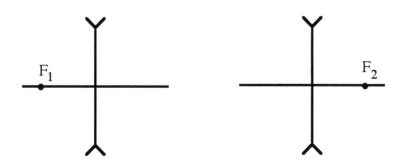

10.4 Point F_2 is a principal focus of the diverging lens, which intercepts a cone of rays converging toward point F_2 from the left.

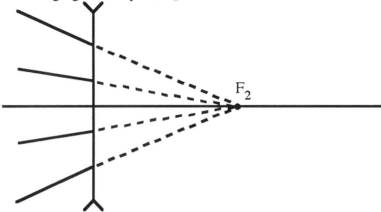

Add to the diagram and label the following additional lines:

(a) A wave front associated with the incoming converging rays.

(b) The rays that emerge from the lens on the right-hand side.

(c) A wave front associated with the emerging rays.

10.5 A thin converging lens L has principal foci at F_1 and F_2. The lens forms an image (not shown) of object O.

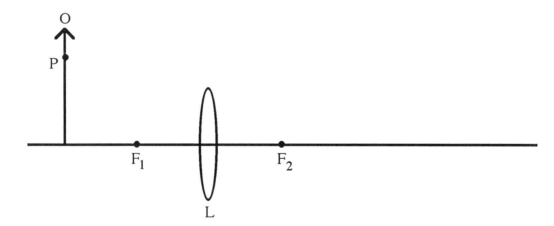

(a) Is the image formed by the lens that of the *entire* object or does the image encompass only *part* of the object? Explain your reasoning.

(b) Consider the point P. Like any other point on the object, this point sends out light rays in all directions. Determine what happens to light emanating from point P: What happens to the rays that are not intercepted by the lens? What happens to the rays that *are* intercepted by the lens? Give the answers by drawing appropriate rays on the diagram, and then describe the results in words.

(c) Sketch a few wave fronts for rays from point P both before and after interception by the lens.

10.6 A thin converging lens with principal foci at F_1 and F_2 forms a real image I at the position shown in the diagram. Draw the three principal rays that will permit you to determine where the object must have been in this case, and describe in words how each ray is drawn.

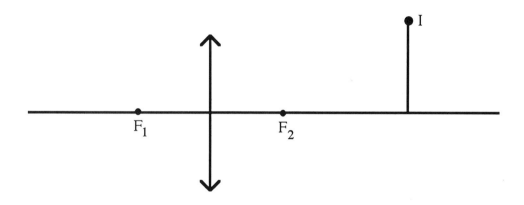

10.7 A thin converging lens forms a virtual image I at the position shown in the diagram. Draw the three principal rays that determine where the object must have been in this case, and describe in words how each ray is drawn.

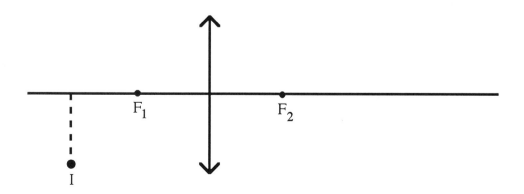

10.8 The dashed arrow and symbol O mark the location of a real image that is being formed by a converging lens some distance off to the left. The light coming from the converging lens is intercepted by the diverging lens shown in the figure. The arrow O is therefore a *virtual* object for the diverging lens, the principal foci of which are located at the points labeled F_1 and F_2, respectively.

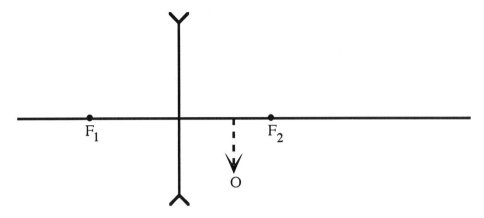

(a) Draw the three principal rays so as to establish the location and size of the final image.

(b) Describe in your own words how each of the three principal rays is drawn.

(c) Indicate whether the final image is real or virtual and explain how you arrived at your conclusion.

(d) Sketch a few wave fronts for the rays intercepted by the diverging lens and for the rays emerging from the diverging lens.

10.9 A thin converging lens is placed at position A on an optical bench as shown in the *upper* part of the following diagram. The lens has a focal length of 3.00 cm, and its principal foci are shown at the positions marked A_1 and A_2, respectively. An object O_1 is placed at a position 4.00 cm to the left of the lens. (The diagram is drawn to scale.)

(a) On the upper part of the diagram, draw the principal rays that locate the image formed by the converging lens. Describe in your own words how each ray is drawn (put this writing elsewhere on the page; do not clutter the diagram).

(b) Use the relevant lens equations to calculate the position of the image and the lateral magnification.

c) Interpret the results you obtained in parts (a) and (b): Do your graphical and algebraic results agree or disagree? Is the image real or virtual?

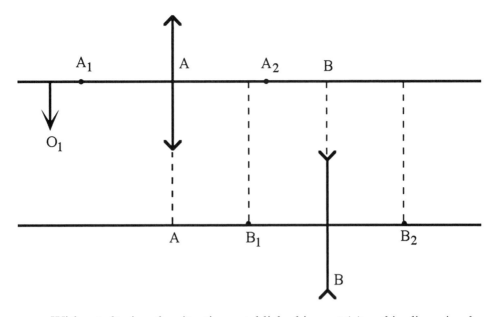

Without altering the situation established in part (a), a thin diverging lens is now placed at position B on the optical bench just 5.00 cm to the right of lens A, intercepting the cone of light coming from lens A. The focal length of lens B is 2.50 cm, and the principal foci are marked B_1 and B_2. (The diverging lens is shown in the *lower* part of the diagram, but this is done to avoid cluttering the upper part. The two lenses are actually in line on the same optical bench.) The image formed by lens A now becomes the object for lens B.

(d) Is the object for lens B real or virtual? Explain your answer.

(e) Complete the diagram, drawing the three principal rays establishing the final image formed by lens B. Describe in your own words how each ray is drawn. (If you use a second color for the principal rays of the diverging lens, you may complete the diagram in the upper portion where you have already established the image formed by lens A. If you are using a single color, complete the drawing in the lower portion of the diagram.)

(f) Use the lens equation to calculate the position of the final image and the lateral magnification produced by lens B.

(g) Interpret the final results: Is the final image real or virtual? Do your graphical and algebraic results agree or disagree?

10.10 In this system of object O and converging lens L, the plane mirror M reflects the light emerging from the lens.

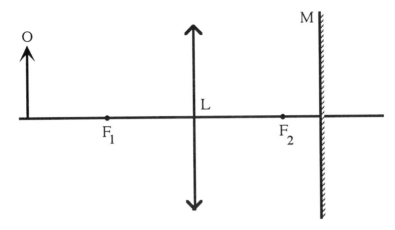

(a) Carefully draw a ray diagram establishing the final location of the image being formed by the system. Is the object for the *mirror* real or virtual? Explain your reasoning.

(b) Is the final image real or virtual? If it is real, how do you reconcile this with the fact that the images you normally see in a plane mirror are always virtual?

(c) Describe what you might do with either the position of object O or the position of the mirror to make the object for the mirror real instead of virtual. Where would the final image formed by the system then be located?

10.11 Point O is a source of light. Two rays from O are shown passing through the thin converging lens and crossing each other at point C.

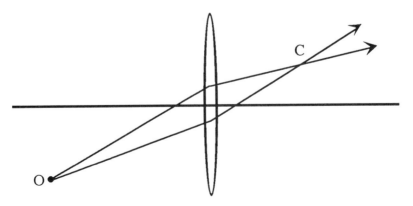

(a) Find the two principal foci of the lens by drawing appropriate rays, and label them F_1 and F_2, respectively. Explain your reasoning.

(b) Draw a ray from O to the extreme upper end of the lens and show the direction in which it emerges from the lens. Explain your reasoning.

(c) Where would you place a screen so that the image of O would be in focus on it? Explain your reasoning.

(d) Show on the diagram where you would place your eye, and the direction in which you would look, to see the image of O without the presence of a screen. Explain your reasoning.

(e) Suppose the lens is replaced by another lens having the same focal length but a larger diameter. Will the location of the image along the principal axis change? Why or why not? If it changes, how will it change? Will the distance of the image from the principal axis change? Why or why not? If it changes, how will it change? Will the brightness of the image change? Why or why not? If it changes, how will it change?

10.12 Consider the following statement: "If it were possible to make perfectly smooth, perfectly spherical lenses, one would then have lenses that would not have to be corrected for aberrations." Is this statement true or false? Indicate your line of reasoning.

10.13 Consider the following statement: "In using a slide projector in which a large image is projected on the screen, the slide must be located just inside the principal focus of the lens (i.e., slightly closer to the lens than the focal point)." Is this statement true or false? Explain your reasoning with the aid of an appropriate diagram.

10.14 Consider the following statement: "When a camera is used to take pictures, the film must be located at the principal focal plane of the lens for very distant subjects and somewhat closer to the lens than the principal focal plane for closer subjects." Is this statement true or false? Explain your reasoning with the aid of an appropriate diagram.

10.15 Consider the following statement: "As we come closer to an object, the lens system of our eye must become more strongly converging, to focus the image sharply on the retina." Is this statement true or false? Explain your reasoning with the aid of an appropriate diagram.

10.16 Let us examine a very basic aspect of the everyday use of cameras and slide projectors.

(a) Suppose you are using a camera and wish to have a larger image of a distant object than you are obtaining with the lens currently in use. Would you change to a lens with a longer or a shorter focal length? Explain your reasoning.

(b) Suppose you are using a slide projector and wish to obtain a larger image on the screen. You cannot achieve this by moving the screen farther from the projector because you are already using the entire length of the room. Would you change to a lens with a longer or a shorter focal length than the one you are using? Explain your reasoning.

It is possible to address these two questions simply by drawing and interpreting appropriate geometrical diagrams. It is a valuable exercise, however, also to address the questions mathematically (by appealing to the lens equation and the expression for lateral magnification) and proving your contentions analytically. This is a nice exercise in elementary mathematical physics. Be sure to verify that your analytical solution agrees with your geometrical one. [Hint: It is important to realize and keep in mind that in part (a) the object distance is essentially fixed while in part (b) the image distance is essentially fixed.]

Note to the student: In the following multiple-choice questions, mark those statements that are correct. Any number of statements may be correct, not just one. You must consider each statement on its merits.

10.17 A simple glass lens exhibits what is called ''chromatic aberration''; i.e., its focal length for red light is longer than its focal length for blue. This behavior of the lens can be explained on the basis that

(a) the resolving power of a lens is greater for blue light than for red.

(b) longer wavelength rays are refracted more strongly at the edges of the lens than nearer the center.

(c) shorter wavelength rays are refracted more strongly at the edges of the lens than nearer the center.

(d) red light is retarded less strongly than blue on passing from air to glass.

(e) red light is retarded more strongly than blue on passing from air to glass.

(f) the diffraction introduced by the lens separates the colors.

(g) the lens is not perfectly spherical.

(h) the lens surface is not as smooth as is necessary for perfect focussing.

(i) None of the above.

10.18 This diverging lens has principal foci at F_1 and F_2. A ray of light AB is incident from the left at point B on the lens.

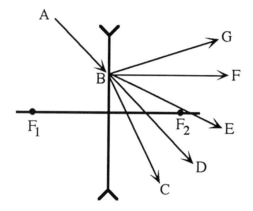

The emerging ray is best represented by

(a) ray BC.

(b) ray BD.

(c) ray BE.

(d) ray BF.

(e) ray BG.

Explain your reasoning.

10.19 The arrows in the following diagram represent object and image formed by a diverging lens. Which of the points marked A, B, C, D, E is closest to a focal point of the lens?

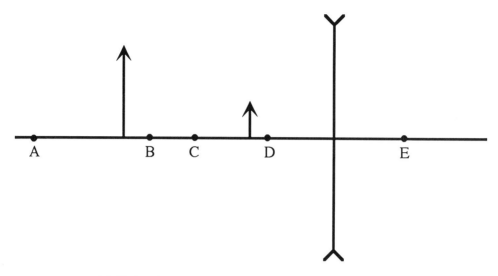

(a) Point A.

(b) Point B.

(c) Point C.

(d) Point D.

(e) Point E.

(f) Both points D and E.

Explain your reasoning.

CHAPTER 11

Geometrical and Physical Optics

11.1 In diagrams (a) and (b), a bright source of light casts the shadow of an opaque obstacle B on a wall that is a diffuse reflector of light (not a mirror). The eye of an observer is shown at E. In (a) the source of light, P, is to be treated as a point source of negligible spatial extent, sending out light in all directions. In (b) the source of light, S, also sends out light in all directions, but the source has substantial size, as shown.

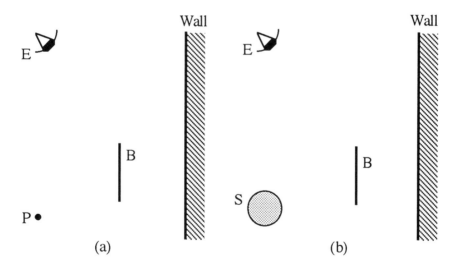

(a) In each diagram, draw rays of light that originate at the source and reach the eye of the observer, setting the boundaries of the shadow that the observer sees on the wall. Interpret the diagrams so as to explain how it comes about that the shadow in (a) exhibits only one region, uniform in darkness, while the shadow in (b) consists of two regions, a dark inner region (called the "umbra") and a lighter outer region (called the "penumbra").

(b) When the shadow of an object is cast by the sun, the shadow is quite sharp when B is relatively close to the wall and is much fuzzier, with obvious umbra and penumbra, when B is relatively far from the wall. How do you explain the difference between these two situations?

173

11.2 In an illuminated room, you can see any wall of the room from any location in which you stand.

(a) Is there any reflection involved in the interaction between light and the walls? If not, how do you account for being able to see the wall at all? If so, how do account for the fact that you do not see reflections of other objects in the room?

11.3 If we see a reflection of the moon or the sun in a very still lake, we see a relatively sharp outline with perhaps a few wiggles around its periphery. If the water surface is ruffled, however, we see a broad streak stretching toward us along the surface.

(a) Explain these observations in terms of what you now know about the reflection of light. Use sketches to assist and clarify your explanation.

11.4 A light ray is incident at an interface between media 1 and 2 and is partially reflected and partially transmitted as shown.

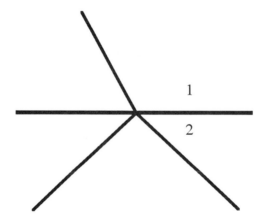

(a) Identify the incident, reflected, and transmitted rays and explain how you made the identification. Draw arrow heads on each ray indicating the direction of propagation of the light.

(b) Label the angles of incidence, reflection, and transmission (refraction).

(c) In which medium does light have the higher velocity of propagation? Explain your reasoning.

(d) Which of the two media has the higher index of refraction? Explain your reasoning.

(e) Draw short lines on each ray showing the orientation of the wave fronts associated with each ray.

(f) Does the frequency of the light change on passing from one medium to the other? If it changes, explain, in terms of what happens at the interface, why it changes and indicate whether it increases or decreases. If it does not change, explain why it doesn't.

(g) Does the wavelength of the light change on passing from one medium to the other? If it changes, explain, in terms of what happens at the interface, why it changes and indicate whether it increases or decreases. If it does not change, explain why it doesn't.

11.5 A layer of water is contained in a glass tank as shown. (The index of refraction of glass is greater than that of water.) A ray of light is incident at the bottom of the tank as shown.

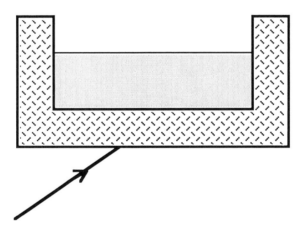

(a) Sketch the ray transmitted into the glass and then into the water and out of the water into the air.

(b) Now assume that the water is removed from the tank. Sketch a ray incident, on the bottom of the tank as before, transmitted into the glass, and then into air instead of water. What differences are there between the rays in parts (a) and (b)? Explain.

11.6 Two very narrow rays of light of different colors are incident on a glass plate at different angles. (The difference in angle is exaggerated for the sake of clarity.) It so happens that the two rays coalesce into one ray on entry into the glass, as shown in the following diagram.

(a) Which of the two rays has the higher index of refraction in glass? Explain your reasoning.

(b) Sketch the rays reflected from the bottom surface of the glass and emerging back into the air through the upper surface of the glass. Explain your reasoning.

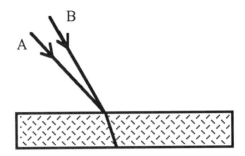

(c) Suppose the angle of incidence of rays A and B were increased without changing the angle between them. At what angle of incidence at the upper surface of the glass would a ray of either A or B, reflected from the bottom surface, undergo total internal reflection within the glass? What would simultaneously be happening to the ray of the other color? Explain your reasoning.

11.7 A hollow ''air prism'' is made with walls of thin transparent plastic and is sealed to be watertight. The prism is immersed in water as shown. A ray of light enters the prism from the water on the right.

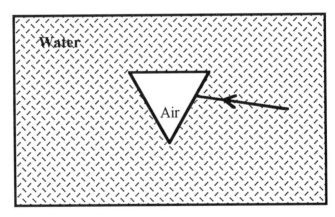

(a) Sketch a possible refracted ray entering the prism, passing through it, and emerging back into the water through another wall. Show normals to the surface and reflected rays as well the refracted ray at each interface. (The walls of the prism are to be treated as having negligible thickness and playing no significant role in the path of refraction.)

11.8 A lens of glass or transparent plastic is molded into the shape shown in the figure. A parallel beam of light, parallel to the principal axis, is incident on this lens from the left .

(a) Will the beam emerging from the lens be converged or diverged, or will it remain parallel to the principal axis? Explain your reasoning.

11.9 In the following diagrams, region I consists of air with an index of refraction of 1.0, and region II consists of a solid transparent material with index 2.0. The sides of the transparent solid are perpendicular to each other at the vertex, and the dashed principal axis bisects this right angle. In the first diagram, a point source S, sending out light in all directions, is located on the axis in region I, and, in the second diagram, it is located on the axis in region II. (In the following questions, confine your investigations to the plane represented in the diagrams, i.e., the plane of the paper.)

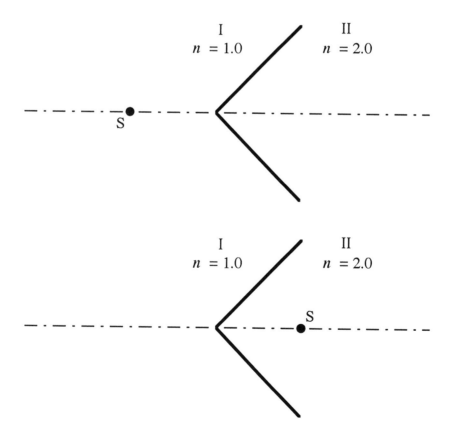

(a) In the upper diagram, is there a region within region II in which a detector would "see" two apparent point sources of light? If you conclude there is no such region, explain your reasoning. If you conclude there *is* such a region, map out its boundaries in the plane of the diagram, explaining how you arrived at this conclusion and how you established the boundaries.

(b) In the lower diagram, is there a region within region I in which an observer would see two apparent sources of light? If you conclude there is no such region, explain your reasoning. If you conclude there *is* such a region, map out its boundaries in the plane of the diagram, explaining how you arrive at this conclusion and how you establish the boundaries.

11.10 A thin wedge-shaped film of air is present between two glass plates as shown in the following diagram. The thickness of the film is essentially zero where the plates are in contact at the bottom of the figure, and the thickness at the top has the unknown value denoted by δ. The plates are illuminated from the left by a parallel beam of monochromatic light with a wavelength of 5400 Å. When the reflected light is viewed from the left, one sees the interference pattern shown at the left.

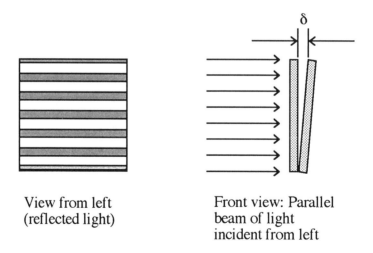

View from left
(reflected light)

Front view: Parallel
beam of light
incident from left

(a) What is the color of the light being used? (Cite your source of information.)

(b) Given the nature of the observed pattern, what can you say about the flatness and smoothness (or the departure from flatness and smoothness) of the surfaces of the two plates? What might be the nature of the pattern if the surfaces were not very flat? Explain your reasoning.

(c) Calculate the thickness δ of the air film at its upper end, explaining your reasoning and giving the result in both angstrom units and centimeters.

11.11 A thin film of oil clings to the left-hand side of a very flat and smooth glass plate, as shown. Under the influence of gravity, the oil film takes on a wedgelike shape, since it tends to thicken at the bottom. The thickness of the film is essentially zero at the top and 9750 Å at the bottom. The indices of refraction of the oil and the glass are 1.4 and 1.6, respectively.

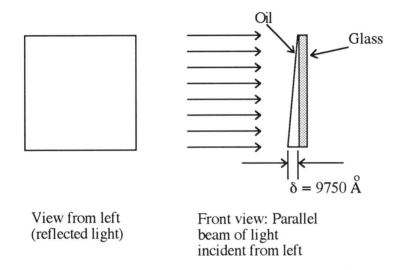

Oil

Glass

$\delta = 9750 \overset{\circ}{A}$

View from left
(reflected light)

Front view: Parallel
beam of light
incident from left

A parallel beam of red light (wavelength 6500 Å) is incident from the left. The film is viewed from the left, and the reflected light forms an interference pattern. Determine the location and the number of bright and dark bands that will be observed in the interference pattern and sketch the predicted pattern in the blank square at the left-hand side of the figure. Be sure to explain your reasoning and mark which bands are bright and which are dark.

11.12 Consider the following statement: "As an optical grating is made finer (i.e., as spacing between lines is *decreased*), the various orders of spectra (1st, 2nd, 3rd, etc.) all lie closer to the principal axis." Is this statement true or false? Explain your reasoning.

11.13 Consider the following statement: "The fact that a lens is capable of producing a real image by refracting light and focussing it on a screen demonstrates that light must be a wavelike phenomenon." Is this statement true or false? Explain your reasoning.

11.14 A narrow beam of white light is spread out into a spectrum on passage through a grating and on passage through a prism.

 (a) Sketch the spectrum that would be observed in the case of the grating with emphasis on the location of the red and blue ends of the spectrum with respect to the principal axis of the system.

 (b) Make a similar sketch for the case of the spectrum produced by the prism.

 (c) State clearly and concisely in your own words what *inferences* are to be drawn from these observations about comparative properties of red and blue light. That is, what does each experiment tell us about properties such as wavelength and/or velocity of propagation of the different colors? What does each experiment *not* tell us about wavelength and/or velocity of propagation of the different colors?

11.15 Suppose a glass (index of refraction 1.52) converging lens with a given focal length is placed in water (index of refraction 1.33). What do you predict will happen to the focal length of the lens while it is located in the water? Will it increase, decrease, or remain unchanged? Explain your reasoning.

11.16 In the so-called Newton's rings experiment, a collimated (parallel) beam of monochromatic light is incident on the system illustrated: a plano–convex glass lens (A) resting against a glass plate (B) with air between the two pieces of glass. A and B are made of the same kind of glass. Both the light reflected from this system and the light transmitted through it form patterns consisting of concentric bright and dark rings. We shall concentrate on the reflection pattern. This pattern begins with a *dark* circular region at the center, followed by a next bright ring, etc. (If you have not seen a Newton's rings demonstration, you should ask to see one.)

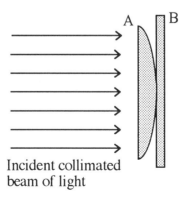

Incident collimated
beam of light

Early in the nineteenth century, a controversy raged over the comparative merits of the wave and particle models of light. Opponents of the wave model pointed to the dark central region of the Newton's rings reflection pattern as a serious failure of the model. Their reasoning was that since the air film at the center, where the two glasses are in "contact," must be extremely thin relative to the supposed wavelength of the light, there should be zero phase shift between the light reflected from the lens-air interface and that reflected from the air-plate interface, since the back and forth travel time in the film would be essentially zero. If interference is taking place between the two reflected beams, it should be constructive rather than destructive, and the central region of the pattern should be bright rather than dark.

It occurred to Thomas Young, vigorous proponent of the wave model, that by analogy to the behavior of waves in mechanical systems, light waves might undergo either a zero or a complete 180° phase shift on reflection at interfaces, depending on whether the index of refraction increases or decreases at the interface. (Young conceived of change of index as analogous to relatively fixed or relatively free boundaries for waves on strings or springs.)

Young proceeded to alter the setup illustrated above by making lens A out of crown glass with an index of refraction of about 1.5 and the plate B out of flint glass with an index of about 1.7. Instead of air between the two pieces of glass, he put a film of sassafras oil, having an index of about 1.6. The pattern on reflection now showed a bright, instead of a dark, central region. The experiment proved to be a compelling victory for the wave model.

(a) Interpret Young's clever experiment in your own words. What was the point of using the sassafras oil? How did the experiment support Young's hypothesis concerning phase shifts? What information does it *not* give concerning the phase shifts? How does the overall combination of observations, including the earlier ones, support the idea that the observed patterns are interference patterns and thus support the wave model?

(b) Does Young's experiment give any direct information about "what is waving" in a "light wave"? Does it give any information as to whether the wave is longitudinal or transverse? Explain your reasoning.

(c) In the foregoing sequence of experiment and reasoning, identify the observations and the inferences drawn from the observations, distinguishing clearly between the two modes. What information that might be desired *cannot* be drawn from these specific observations?

(d) Why do you think it was customary in the 1800's to refer to media with higher indices of refraction as optically "more dense" than media with lower indices?

11.17 When a narrow, circular beam (pencil) of monochromatic light is incident on a diffraction grating whose lines run *vertically*, the resulting diffraction pattern (to the first order) is that shown in figure (a) on the left. If the lines of the grating are oriented *horizontally*, one obtains the pattern shown in (b) on the right.

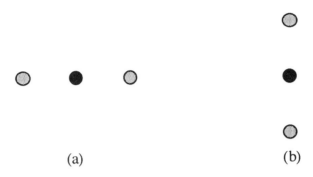

(a)   (b)

If one now *crosses* the two gratings one on top of the other (i.e., the lines of one run horizontally while those of the other run vertically), one obtains the following pattern, with nine spots instead of six in the first-order spectrum.

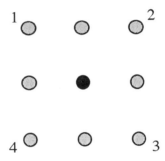

(a) How do you account for the sudden appearance of the four new spots (1, 2, 3, 4) on the diagonals at the corners of the square, spots that were not present at all with the individual gratings? [Hint: Draw a picture of the overlapping grating lines, and examine the pattern of openings carefully for lines of openings (other than horizontal and vertical) that would lead to diffraction patterns.]

(b) Note the relative *spacing* between the array of openings that you identify as causing the new spots and show that these spots *must* lie exactly on the corners of the square.

(c) Suppose you rotated the crossed gratings rapidly around the central axis in the plane of the paper. What would be the resulting pattern of bright and dark areas?

11.18 Suppose a monochromatic point source of light P is placed very close to a good plane mirror as shown in the following diagram. The result is an interference pattern in the region above the mirror, the effect being known as "Lloyd's Mirror."

P
o
_____

(a) Explain the origin of the interference pattern, making a sketch of wave fronts to show how the pattern is developed in the plane of the diagram. What is the independent variable that determines the angular spread of the pattern? How is this situation connected with the two-slit interference pattern you have already studied? Explain.

(b) Making use of results you have already derived in class or textbook, write equations for the angular location of constructive and destructive interference loci in the plane of diagram. The angular positions should be expressed in terms of the relevant independent variable you defined in part (a).

(c) Describe how you might set up an analogous situation in a ripple tank.

CHAPTER 12

Fluids and Thermal Phenomena

12.1 A circular cylinder, open to the atmosphere, contains 2500 cm³ of a liquid having a mass of 0.87 g in each cubic centimeter. The atmospheric pressure is 1.00 bar. (Be sure to explain your reasoning in each of the following calculations.)

(a) Calculate the total *mass* of the liquid in SI units.

(b) Calculate the total weight of the liquid in SI units.

(c) If the height of the liquid in the cylinder is 120 cm,

  (1) Calculate the gauge pressure p_g at the bottom of the cylinder.

  (2) Calculate the absolute pressure p_{abs} at the bottom of the cylinder.

(d) If all the liquid is poured into a second cylinder with a diameter 1.80 times larger than that of the first cylinder, calculate the ratio of the gauge pressure p_{2g} on the bottom of the second cylinder to the gauge pressure p_g previously exerted on the bottom of the first cylinder. (Solve this problem by ratio reasoning, *not* by substitution in formulas.)

(e) Is the ratio of absolute pressures at the bottoms of the cylinders in part (d) the same as the ratio of the gauge pressures? Why or why not? Explain your reasoning.

12.2 A container of water rests on a platform scale as shown, and the needle gives the reading W for the weight being measured.

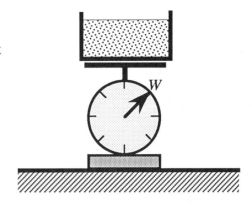

(a) How does the reading W compare with the force exerted by the water on the bottom of its container plus the weight of the container, i.e., is W equal to, greater than, or less than this combined force? Explain your reasoning.

(b) Does the reading W include the weight of the column of air above the container? Why or why not?

(c) Would you expect the reading W to change significantly if the atmospheric pressure were decreased? Why or why not? If you expect it to change, would it increase or decrease? Explain your reasoning.

12.3 Consider two rectangular tanks of water: the side walls (left and right) of the tanks have exactly the same area, but the front and back walls do not because the lengths of the tanks are quite different. Both tanks are filled to the same depth and obviously contain very different total weights of water.

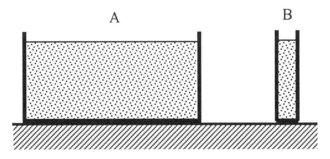

(a) How does the total outward force on the left and right side walls of container A compare with the total outward force on the left and right side walls of container B? Are the forces equal or is one greater than the other? Explain your reasoning.

(b) How do the total outward forces on the front and back walls of container A compare with the corresponding forces in container B? Explain your reasoning.

12.4 The diagram represents a container filled with liquid of density ρ. In our imagination we shall "extract" from the interior of the liquid the two columns labeled A and B. Let us denote the cross-sectional (or base) areas of the two columns by the symbols ΔS_A and ΔS_B and their heights by y_A and y_B, respectively. We denote the atmospheric pressure by p_o.

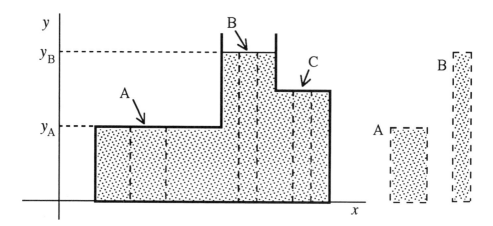

(a) Draw free body force diagrams for columns A and B on the right, showing all the *vertical* forces being exerted on each. (To avoid cluttering the diagram, do not try to show horizontal forces on the columns even though we know such forces are being exerted.) Describe each vertical force in words. (Verbal description consists of saying what object exerts this force on what.)

(b) In terms of the symbols defined above, what is the algebraic expression for the total force acting on the top surface of column B? On the top surface of column A? Explain your reasoning.

(c) Including expressions for the weight of each column, what must be the expression for the total force acting on the bottom surface of column B? On the bottom surface of column A? Explain your reasoning.

(d) Do your expressions lead to the result that the *pressure* is the same at the base of each column? If not, review your work because you must have made an error in reasoning.

(e) Is the absolute pressure at the bottom of this container equal to the pressure of the atmosphere plus the total weight of the liquid divided by the total area of the base? Why or why not? Under what circumstances, if any, *is* the absolute pressure at the bottom of a container of liquid equal the pressure of the atmosphere plus the total weight of the liquid divided by the total area of the base?

(f) How does the situation at the top of column C compare with that at the top of A and B? Explain your reasoning.

(g) Suppose a hole is made in the wall of the container at the very top of column B. What will happen? Explain your reasoning. Suppose a hole is made at the top of column A?

12.5 An apocryphal story goes as follows: A large van stops at a traffic light, and the driver leans out and pounds on the wall with a baseball bat. A puzzled driver in an adjacent lane asks what is going on. The van driver replies, "This is only a five-ton van, and I am carrying eight tons of canaries, so I have to keep at least three tons of them flying."

(a) Assess the van driver's physics. Can he get away with this strategy? Why or why not? What must actually be happening inside the van when the canaries are flying? What is the force on the floor of the van regardless of whether the canaries are roosting or flying? Explain your reasoning.

12.6 Consider the following barometer-like arrangement. Atmospheric pressure has its normal value corresponding to 76 cm of mercury, but the column of mercury in the tube stands at a height of $y = 60$ cm instead of 76 cm above the surface of the mercury in the reservoir, and there is a space between the top of the mercury column and the top of the tube.

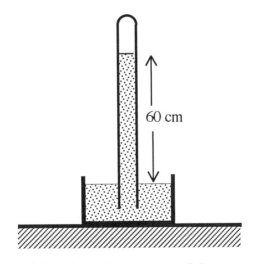

(a) What do you infer about the contents of the space in the tube above the top of the mercury column compared with the contents of this space if the column height were 76 cm? Explain your reasoning.

(b) How does the pressure in this space compare with atmospheric pressure: Is it equal to, greater than, or less than atmospheric pressure? Calculate the ratio of any pressure in the space to atmospheric pressure. Explain your reasoning.

(c) What would happen to the height y of the mercury column if the outer container were very much deeper and more mercury were poured into the reservoir, raising its height in the outer container? Would y increase, decrease, or remain unchanged? What would simultaneously happen to the pressure in the space above the top of the column? Contrast this result with what would happen if the space above the column were a perfect vacuum. Explain your reasoning.

12.7 Let us imagine an iceberg as a neatly rectangular block of ice floating in seawater. The density of sea ice, with its inclusions of air bubbles, snow, and drops of liquid water, is about 0.89 g/cm^3. The density of seawater is about 1.03 g/cm^3.

(a) Draw a force diagram of the floating block, representing the distributed force exerted by the surrounding water by a single total force. Describe each force in words.

(b) Stating the appropriate governing principle or principles and explaining each step, calculate the *fraction* of the height H of the block that will stick up *above* the water surface.

(c) Consider a second block of sea ice having the same density as the one above but scaled down in size, with all three length dimensions being half those of the original. What *fraction* of the height of this second block will protrude above the water surface? Be sure to explain your reasoning.

12.8 If a tank of water is accelerated to the right (in a car or on a cart), the water shifts to the left, and the surface takes on a slope downward to the right as shown in the following diagram. A constant downward slope is maintained as long as the acceleration is constant.

(a) Every particle within the body of water is being accelerated to the right, along with the entire system. Draw separate free body force diagrams for the parcel of fluid in box A and the parcel of fluid in box B, showing larger forces with longer arrows and equal forces with arrows of equal length. (Use single arrows for forces on the sides of each box, even though the force is actually continuously distributed; this is what we usually do, for example, with normal, gravitational, and frictional forces, which are also continuously distributed.) Explain any reasoning behind the drawing of unequal forces.

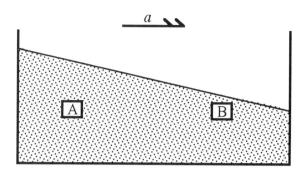

(b) Use your force diagrams to explain why each of the parcels of fluid has the same horizontal acceleration and zero vertical acceleration even though they are located at unequal depths within the fluid.

(c) Suppose a pendulum bob is suspended from the roof of the same accelerating car. How would you expect the angle θ between the pendulum string and the vertical to compare with the angle between the water surface and the horizontal? Explain your reasoning.

12.9 A cylinder of water fastened to a string is whirled around at constant angular velocity in a circle lying in very nearly a horizontal plane. The diagram shows an instantaneous view of the cylinder from the side when its axis happens to lie in the plane defined by the paper.

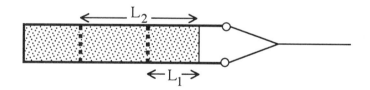

(a) Draw separate free body force diagrams for the two columns of water of lengths L_1 and L_2, respectively. (Draw longer arrows for larger forces and arrows of equal length for forces of equal magnitude.)

(b) Explain how your diagrams account for the centripetal forces necessary to keep all the water in the tube moving in a circle.

(c) Argue that what you have said in parts (a) and (b) means that there must be a pressure gradient (continuously increasing pressure) from right to left in the fluid. Compare this with the pressure gradient that exists in the fluid when the cylinder simply stands upright in the laboratory.

(d) Suppose a small cork is floated in the cylinder before the system is put into revolution. Where will the cork finally be located in the fluid in the steadily revolving cylinder? Explain your reasoning.

(e) Suppose the cylinder is whirled around in a *vertical* instead of a horizontal circle. Again by drawing force diagrams, examine the forces acting on the columns L_1 and L_2 at the instants the cylinder is at the top of the circle, at the bottom of the circle, and halfway in between (i.e., at the instant the string is horizontal). Describe the differences and similarities between each of these three cases on the one hand and the situation in the horizontal circle on the other.

(f) When the cylinder is revolved in a vertical circle, under what circumstances will the liquid remain in the cylinder when the latter is at the top of the circle and under what circumstances will the liquid tend to fall out? Explain your reasoning.

12.10 Consider the following objects: a thin brass ring and a brass plate (of the same composition as the ring) with a hole in it. The two objects start at the same temperature, and the temperature of each is then increased slowly and uniformly (i.e., the temperature does not vary *within* each object).

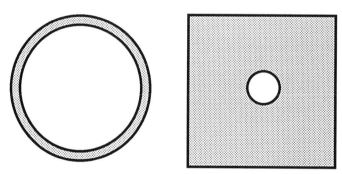

(a) Discuss what will happen to the dimensions of each object, i.e., to the inner and outer diameters of the ring, and to the lengths of the outer edges of the plate and the diameter of the hole in the plate. We are not concerned with numbers but only with directions of change: increase, decrease, or no change. Explain your reasoning carefully, not only to defend your conclusions, but also to be able to refute conclusions that differ from yours. Give especially careful attention to what happens to the inner diameter of the ring and to the diameter of the hole in the plate.

12.11 The coefficient of thermal expansion β of liquid water varies appreciably with temperature at fixed atmospheric pressure. Experimental measurements of β between 10 and 20 °C, at atmospheric pressure, are fairly well represented by the following empirical relation [where $\beta(t)$ indicates that β is a function of the Celsius temperature t]:

$$\beta(t) = 87.9\left(10^{-6}\right) + 11.9\left(10^{-6}\right)(t - 10)$$

(a) Define the term "empirical" in the foregoing context.

(b) Define the term "coefficient of thermal expansion."

(c) Making appropriate use of the calculus and explaining each step of reasoning, calculate the *change* in volume that takes place when 1500 cm^3 of liquid water increase in temperature from 10 °C to 20 °C at constant atmospheric pressure.

d) Sketch a graph of $\beta(t)$ versus t, and interpret the change in volume you have calculated in part (b) as a quantity related to some property or feature of this graph. How is this graph related to a graph of volume versus temperature?

(e) In terms of what you have said and done in parts (b) and (c), explain, as though you were dealing with an uncertain fellow student, why it is *incorrect* to say that the volume change can be found by simply multiplying out 1500 × (87.9) × (10^{-6}) × (10). Connect your explanation with the graphs sketched in part (d) as well as with the arithmetic involved.

12.12 Although liquid-in-glass thermometers are usually made of glass that expands very little on increase in temperature, some expansion of the glass does occur in most instances, especially for fairly large temperature changes.

(a) As the glass expands, what happens to the volume of the inner space occupied by the liquid? Does it increase or decrease? Explain your reasoning.

(b) How do you explain the fact that the liquid does rise in the thermometer with increasing temperature even though the glass expands? Cite your evidence and explain your reasoning carefully.

(c) Do you expect the scale of graduations on a thermometer to be perfectly uniform over the entire length of the instrument? Why or why not?

12.13 (1) It is an observed fact that liquids and solids can be compressed (i.e., made smaller by application of external forces), but it takes very, very large forces to produce very small changes in volume (or length). (2) You are certainly aware that bottles (or water pipes) burst when water that fills them freezes.

(a) What information does fact 2 give us about the behavior of the volume of a parcel of water when it freezes? Explain your reasoning.

(b) What connection do you see between fact 1 and the bursting described in fact 2? Explain your reasoning.

12.14 A metal rod is held firmly within the jaws of an iron clamp. A blowtorch is now directed at the metal rod (not the clamp), and the temperature of the rod is increased without appreciable change of the temperature of the clamp.

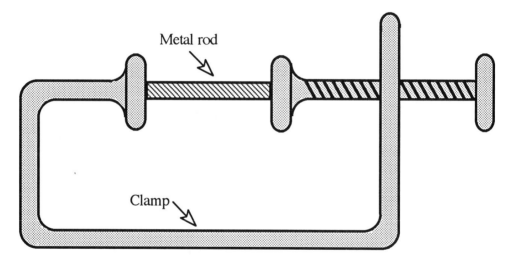

(a) What is likely to happen to the clamp as the temperature of the rod increases? Explain your reasoning. How is the effect you are describing connected with our awareness that although solids can be compressed (made smaller in size) by application of external forces, very large forces are required to produce very small changes in dimensions?

12.15 It is a matter of everyday experience that when objects at different temperatures are brought in contact, an "interaction" takes place. The temperature of each object changes until a stopping point (equilibrium) is reached. We call this a "thermal" interaction. (We recognize, in the world around us, many other kinds of interaction in which objects, or groups of objects "do something" to each other. We speak of mechanical, gravitational, electrical, magnetic, chemical interactions, etc.) Let us reexamine our everyday experiences to pull out the regularities in thermal interactions that remain hidden unless we think about them.

(a) We have two containers of water, each with its own thermometer. What happens if we bring the two, one at higher and one at lower temperature, in contact with each other, either by simply mixing the waters or by good, direct contact between the containers? (If we imagine doing such an experiment by direct contact between containers, it is well to imagine separating the two containers from contact with air in the room by keeping them in a good, thick box.) In which direction (increase or decrease) does each thermometer change when contact begins? At what point as far as the thermometers are concerned is equilibrium reached? When, that is, do changes cease? If the two amounts of water are different, which ends up with the bigger temperature change when equilibrium is attained? If you made one of the water quantities very, very much larger than the other, where would the temperatures end up? How would the temperature change of the huge body compare with that of the tiny one? Is the temperature change of the large body ever actually zero?

(b) What is the end point of temperature changes for a container of hot water put out in a room that is initially at room temperature (20 °C)? What bodies undergo temperature changes in the interaction? What would happen to the temperature of the air in the room if the water container were very large? Answer the same questions for the case in which the container of water is very cold instead of very hot. How do these situations relate to the ones you discussed in part (a)? How is the temperature of a room maintained at a comfortable level during cold weather?

(c) Placing materials such as wood, glass, or fiber between thermally interacting objects greatly slows down the temperature changes but does not make them stop at unequal temperatures. We speak of such layers of material as providing ''thermal insulation.'' Explain, in the light of your discussions in parts (a) and (b), the point of thermally insulating a house and the point of using containers that tend to keep cold drinks cold.

(d) Have you ever seen a case in which a colder object gets colder and a hot one hotter on contact? Or a case in which two objects start at the same temperature and one spontaneously increases in temperature while the other decreases? Why would it be exceedingly desirable to be able to make a kettle of water boil by bringing it into contact with an object at the same initial temperature? It is perfectly possible to *imagine* such changes, but do you believe them to be possible? Why or why not?

(e) What generalizations about thermal phenomena can you now put together in the light of visualizing and relating the everyday experiences we have been describing?

12.16 We tend to use the word "heat" very casually in everyday speech, and many individuals use the words "temperature" and "heat" synonymously; i.e., they do not distinguish the terms as having very different meanings. In physics, "heat" is a technical term and denotes an abstract *concept* that is *defined* when the need is recognized through experience. Let us, in this question, retrace some of this experience. We start only with the idea of "temperature" as the reading on the familiar thermometer. When we examine some common thermal interactions, we begin to find that the thermometer does not tell us the whole story concerning the interactions (if it did, we would not need to invent an additional concept).

(a) Consider three identical quantities of hot water, all at exactly the same initial temperature. We put them out in a room at a fixed, lower temperature, but we use three different containers. The first sample is contained in an ordinary glass with no wrapping. The second sample is placed in a similar glass, but the glass is wrapped with a thick layer of plastic material such as styrofoam. The third sample is placed in a high quality thermos bottle. Note carefully what your experience tells you: all three samples start at exactly the same initial high temperature and end up at exactly the same (lower) room temperature. So there is no difference between the initial temperatures and no difference between the final temperature readings when equilibrium is attained. There is, however, a very significant difference between the intermediate histories of the three samples. Describe what actually happens. How do the three interactions differ even though the initial and final temperatures do not differ?

(b) Suppose we start with two very different quantities of water (one larger and one smaller) in separate beakers, initially at exactly the same temperature. We place the beakers on identical heaters and proceed to increase the temperatures in each beaker to the same final value. The temperatures in both samples are the same to begin with and the same at the end; there is no difference among the thermometer readings. Yet there are significant differences between the two interactions. Describe the differences in your own words, being explicit about the time intervals involved and the utilization of fuel or electricity in the heating.

(c) Suppose we start with two equal amounts of water, one sample at room temperature (20 °C) and the other at, say, 80 °C. We immediately mix the two samples. At what temperature do you expect equilibrium to be attained (in absence of significant interaction with the surrounding air in the room)? Explain your reasoning. Suppose the higher temperature sample is made considerably smaller than the lower temperature one? Where does *experience* say the equilibrium temperature ends up? In what way do the thermometer readings alone fail to tell us the whole story of interaction?

(d) If we have a solid material in a test tube that is surrounded by a bath that is high enough in temperature, we might bring the material to its melting point (provided we stay below the melting point of the containers). It is an experimental fact that, if the material in the test tube does melt, the thermometer immersed in this material undergoes *no change* while the solid material is disappearing and begins to show an increase in temperature only after all the material is melted. Explain in your own words why this experimental observation is a *dramatic* illustration of the fact that thermometer readings alone do not tell us the whole story of thermal interaction.

(e) Explain in your own words why the foregoing examinations of familiar thermal phenomena lead us to *invent* a new concept, the concept to which we give the name "transfer of heat."

12.17 Liquid water has a nonzero vapor pressure at *all* temperatures, as confirmed by the fact that it evaporates at any temperature if left standing. If water is evaporating at all temperatures, what is so special about 100 °C? What is it that happens when boiling occurs that is different from what is happening when water evaporates at room temperature?

12.18 If we put some hot water (at initial temperature T_1 at 40 °C) in a beaker and place the beaker in a room where the air temperature has a lower value T_2 at 20 °C, we know that the temperature of the water drops as time goes by, ending up at the same temperature as the air. If we keep track of the water temperature as a function of clock reading (keeping the water well stirred) and graph the data, we obtain a graph roughly like that depicted in the following figure. Such a graph is called a "cooling curve."

(a) Suppose you started with the same amount of water at T_1 but wrapped the beaker in styrofoam (or set the beaker in a tightly fitting styrofoam cup). You now measure the new cooling curve. Sketch what you would expect the curve to look like in comparison with the one sketched in the diagram.

(b) Make similar sketches for the cooling curves you would expect if (1) you put the water in a good thermos bottle and (2) you put the water in a thin metal container.

(c) What is the connection between the sequence of cooling curves you have sketched and the point and purpose of insulating a house in the wintertime? Explain your reasoning.

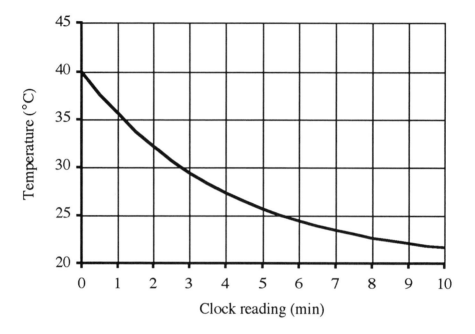

12.19 Joseph Black (1728–1799) was a Scottish physician who, at various times, was professor of medicine and chemistry at the universities of Glasgow and Edinburgh. [Among his students was James Watt (1736–1819), whose improvements and inventions in the design of steam engines played an important role in the advent of the Industrial Revolution in the nineteenth century.] Black was a leader among the researchers who, around the middle of the eighteenth century, refined and sharpened the concepts of ''temperature'' and ''heat.'' Black was probably the first to recognize the significance of the thermal interaction taking place during melting and freezing, when the process of change of phase of a single substance takes place without change in temperature. About this phenomenon, he wrote:

''Melting has been universally considered as produced by the addition of a very small quantity of heat to a solid body, once it has been warmed up to its melting point; and the return of the liquid to the solid state as depending on a very small diminution [of heat]. . . . It was believed that this small addition of heat during melting was needed to produce a small rise in temperature as indicated by a thermometer.

''The opinion I formed is as follows. When ice or any other solid substance is melted . . . a large quantity of heat enters into it . . . without making it apparently warmer as tried by [a thermometer]. . . . I affirm that this large addition of heat [without change in temperature] is the principal and most immediate cause of the liquefaction induced. . . .

"If the common opinion had been well founded—if the complete change of ice and snow into water required only the addition of a very small quantity of heat—the mass, though of considerable size, ought to be all melted within a few minutes or seconds by the heat incessantly communicated from the surrounding air. Were this really the case, the consequences of it would be dreadful . . . for even as things are at present, the melting of large amounts of snow and ice occasions violent torrents and great inundations in the cold countries. . . . But were ice and snow to melt suddenly, as they would if the former opinion of the action of heat . . . were well founded, the torrents and inundations would be incomparably more irresistible and dreadful. . . . This sudden liquefaction does not actually happen. The masses of ice and snow require a long time to melt. . . ."

(a) Discuss Black's argument, translating it into your own words and connecting it with your own experiences. Why do you think Black felt compelled to present this argument to his contemporaries? (Note that what he was doing was creating the then new concept of "latent heat.")

12.20 Following is a schematic (highly idealized) presentation of a temperature versus time graph for an experiment in which a sample of solid ice is taken out of a deep freeze at around –25 °C and heated uniformly in a closed container to a final temperature in excess of 100 °C. The system is closed; i.e., no material enters or escapes from the system. (The presentation is schematic in the sense that real experimental data would show scatter and uncertainty not represented in the graph.)

The following questions pertain to what is happening to the sample during the time intervals indicated by the letters on the graph. For example, during the time interval AB the temperature of the solid ice is increasing with time.

(a) What is happening to the sample during the interval BC?

(b) What is happening to the sample during the interval CD?

(c) What is happening to the sample during the interval DE?

(d) What is happening to the sample during the interval EF?

(e) Under what circumstances is the history denoted by EF possible? What would happen after instant E if the container were open to the atmosphere rather than closed?

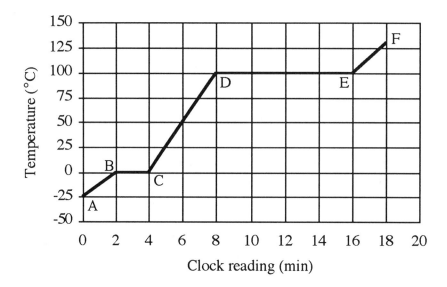

12.21 Joule, in his classic "paddle wheel" experiment (made to determine how much heat, measured in calories, corresponded to a given amount of work measured in mechanical units), used the lowering of a weight to drive a paddle that churned water in a thermally well-insulated container and thus raised the temperature of the water slightly as the mechanical effects were dissipated. (In the following, we shall use modern units, rather than Joule's old English units.)

Suppose that after a block with a mass of 1.80 kg is lowered 16 times through a height of 11.0 m, the 3.00 kg of churned water in the thermally well-insulated bucket exhibits a temperature rise of 0.23 C°. Take the mass of the bucket to be 1200 g, and the specific heat of its material to be 0.107 cal/(g) (C°).

(a) Starting with the block in its elevated position, describe in your own words *all* the energy changes and transformations that take place as the block is lowered.

(b) From the data given above, calculate how many joules of dissipated mechanical work have the same thermal effect on the system as would the transfer of one calorie of heat. (Note that *no* heat was transferred to the water in this experiment; mechanical work was dissipated.)

(c) Suppose the water and the bucket start off at the same temperature as the surrounding air. What effect would a slight thermal interaction with the surroundings (through the insulation on the bucket) have on the result of your calculation? That is, would the interaction tend to make the calculated result higher or lower than the "correct" value? Explain your reasoning carefully on the basis of the equation you set up in making the calculation.

12.22 A plastic tube capped at both ends contains a quantity of lead shot. A common laboratory experiment for determining the number of joules corresponding to one calorie consists of inverting the tube quickly (so that the shot are carried to the top), allowing the shot to fall to the bottom, and repeating the process a number of times. The temperature change of the shot is measured after a recorded number of such falls.

(a) Describe in your own words the sequence of energy changes and transformations that takes place in this experiment. Why does this experiment make it possible to calculate the number of joules corresponding to one calorie?

(b) In an experiment in which the length of the tube was such that the center of mass of the shot fell through a distance of 1.0 m, the initial temperature of the shot was 22.0 °C, and the final temperature, after 50 inversions of the tube, was 25.6 °C. Look up any other data you need in an appropriate table, and calculate what this experiment yields for the number of joules corresponding to one calorie. Be careful about the justified number of significant figures in your final result. Explain all steps of reasoning, and be explicit about any idealizations or approximations introduced in your analysis.

(c) If your result in part (b) differs from the accepted value given in the text, does the deviation make sense? That is, is it in a direction you would expect from the way the experiment is carried out and from the idealizations involved in your calculation? Explain your reasoning.

(d) Why use a plastic tube? Why not a metal tube, for example? Why use lead as the falling material? Why not iron, or copper, or aluminum? (Look up the numerical value of the relevant property of these materials in the appropriate table and make use of this information in responding to the question.) Explain your answers carefully, making reference to the terms that would be affected in the equation you set up in part (b) and inferring the consequences to the accuracy of the experiment. Might the experiment be significantly improved by using a longer tube? Why or why not?

12.23 Several small pieces of copper, having a total (combined) mass of 350 g are placed in liquid nitrogen and, when removed, are at a temperature of − 180 °C. The pieces of copper are quickly transferred to a calorimeter containing 420 g of water at + 9.0 °C. The calorimeter cup is made of brass [specific heat 0.092 cal/(g) (C°)] and has a mass of 200 g. The temperature of the room in which the experiment is conducted is +20 °C.

(a) Describe *qualitatively* what happens in the way of heat transfers and temperature changes after the copper has been placed in the calorimeter. Describe the conservation relation that governs the phenomena taking place, and indicate the idealization we make in applying this relation to the interaction between the copper and the calorimeter. What role does the concept of "closed system" play in the idealizations you invoke?

(b) Now proceed to predict, numerically, the final equilibrium state that is attained within the calorimeter. Explain your reasoning as you set up quantitative expressions and interpret, in words, the physical meaning of each separate term that is present in the equation you end up with. (Hint: You would be wise to make a preliminary, rough calculation to estimate the final physical condition the system attains and to identify the relevant unknown, or unknowns. Otherwise you may find yourself putting numbers into expressions that turn out to be irrelevant.)

(c) Noting that it is, in fact, never possible to attain a perfectly closed system in these circumstances, analyze how interaction with the surroundings will, in this instance, affect the final result: Will the calculated value (of whatever you calculated) be greater than, equal to, or less than the "correct" value that would have been obtained in the ideal situation?

12.24 A refrigerator is left running with its door open in a tightly closed, well-insulated room. As time goes by, what happens to the temperature of the room: Does it increase, decrease, or remain unchanged? Explain your reasoning.

12.25 It is a familiar fact that when electric current is passed through a wire (as in the electric stove, the toaster, the clothes iron, the light bulb, the experiment in the laboratory), the temperature of the wire increases to some point at which the increase ceases. How do you account for the fact that the temperature does not keep increasing indefinitely? What interactions, processes, and effects come into play to stop the increase? What role does the concept of "equilibrium" play in this phenomenon? Explain in as much detail as you can.

12.26 A bicycle pump is full of air at atmospheric pressure; i.e., the piston is as far out in the cylinder as it will go. The stroke of the piston is 14 inches: i.e., the end of the cylinder is 14 inches away from the face of the piston in the initial state. How far must the piston be pushed in before air will begin to enter the tire in which the gauge pressure is 40 lb/in.^{2}? State *all* your assumptions and explain your reasoning. Be careful about the distinction between absolute and gauge pressures and the role they play in your calculation. What would be the effect on your numerical estimate (i.e., Will your estimate be higher or lower than the actual value?) if there were some leakage of air past the piston as compression took place? What assumptions have you made about temperature changes, and how do you justify them? What would be the effect on your numerical estimate of any realistic departure from whatever assumption you have made about temperature changes?

Note to the student: In the following multiple choice questions, circle the letters designating those statements that are <u>correct</u>. <u>Any number</u> of statements may be correct in each question, <u>not</u> just one. You must examine each statement on its merits.

12.27 A cylindrical metal rod M stands on the table. Also standing on the table is a cylinder L containing a liquid. The cylinder is made of a glass called "invar," which does not expand or contract appreciably on change in temperature. Suppose the temperature of the room is increased with corresponding uniform increases in temperature of both systems L and M. Both the metal and the liquid undergo expansion with increase in temperature.

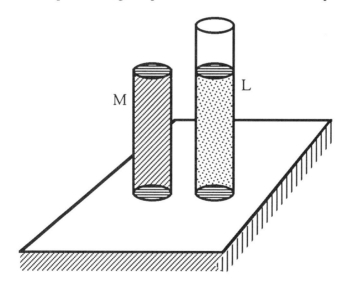

(a) The density both of rod M and of the liquid in L will decrease as the temperature increases.

(b) To calculate the pressure at the bottom of the liquid in L at the new temperature, one would make use of the relation $pV = nRT$, being careful to express the temperature in degrees kelvin.

(c) To calculate the *change* in pressure at the bottom of the rod and at the bottom of the liquid, one would make use of the relation $\Delta p = \rho g \Delta h$, where Δh represents the increase in height that takes place on expansion in each system.

(d) Before the temperature change, the pressure at the bottom of each column would be the same if the heights were the same.

(e) The pressure at the bottom of M will increase as the rod gets longer.

(f) The pressure at the bottom of M will decrease slightly as the rod expands.

(g) The pressure at the bottom of the liquid in L will increase as the temperature increases and the height of the liquid increases.

(h) The pressure at the bottom of the liquid in L will remain unchanged as the temperature changes.

12.28 A flat block of paraffin floats on water with 4.0 mm of its thickness projecting above the water surface. A hole, about 2 cm in diameter is drilled through the center of the block, which is then put back in the water. The block of paraffin now

(a) sinks because of the hole that has been drilled.

(b) floats with its upper surface level with the water surface.

(c) floats with somewhat <u>less</u> than 4.0 mm projecting above the water surface.

(d) floats with somewhat <u>more</u> than 4.0 mm projecting above the water surface.

(e) floats exactly as it did before the hole was drilled.

(f) turns up on edge.

12.29 As water boils at 100 °C at atmospheric pressure, bubbles of gas form in the interior of the liquid and rise to the surface. The bubbles contain primarily

(a) carbon dioxide.

(b) air.

(c) vacuum.

(d) a mixture of the separate gases hydrogen and oxygen.

(e) None of the above.

12.30 Two different samples of the same gas are placed in rigid containers having the same volume. Measurements of pressure and temperature made on the two samples are plotted in the following graph. The data,which show relatively small scatter, are extrapolated by straight lines to zero pressure.

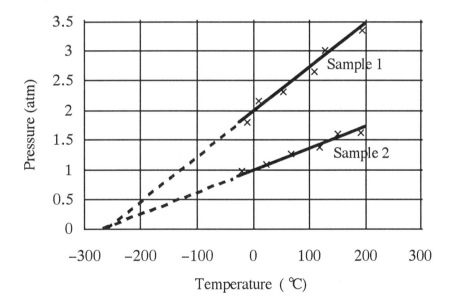

(a) Both samples are exhibiting a close approximation to ideal gas behavior.

(b) Compared with sample 2, sample 1 must contain a larger mass of gas.

(c) Samples 1 and 2 must contain the same mass of gas.

(d) At any given pressure, sample 2 always has the higher temperature.

(e) Some of sample 2 must have been continually leaking away during the measurements.

(f) The two samples disagree with respect to definition of the absolute temperature scale.

(g) The behavior of the two samples points toward definition of the absolute temperature scale.
(h) None of the above.

12.31 The cylinder contains a small amount of liquid water at the bottom. Above the water, the pure water vapor is confined by the piston, which can be raised or lowered. There is *no* air present; the gas phase contains water vapor *only*. The piston walls are made of good heat-conducting material, ensuring that any temperature change within the cylinder is quickly wiped out by conduction to or from the air in the room if the piston is moved quite slowly. Thus the temperature within the cylinder remains very nearly constant. Leakage of gas past the piston is to be considered negligible.

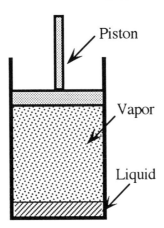

As the piston is moved slowly downward:

(a) the pressure in the cylinder remains unchanged.

(b) the pressure in the cylinder continually increases.

(c) the pressure in the cylinder continually decreases.

(d) the amount of liquid increases while the amount of vapor decreases.

(e) the amount of liquid decreases while the amount of vapor increases.

(f) the amounts of liquid and vapor remain unchanged.

(g) heat flows *into* the system from the surrounding air.

(h) heat flows *out of* the system to the surrounding air.

(i) there is no exchange of heat between the system and the surroundings.

(j) None of the above.

CHAPTER 13

Kinetic Theory

13.1 Once we have accepted the picture of discreteness in the structure of matter (the existence of atoms and molecules as separate entities in the architecture of material substances), we proceed to visualize *gases* as consisting of atoms or molecules that are (1) in continual motion, colliding with each other and with the walls of the container, and (2) on the average, very far apart relative to their own diameters.

(a) What are the justifications for this mental picture or model? In answering this question, draw on what you see in the actual behavior of gases and on what you know about their *macroscopic* physical properties.

(b) Why is this identical model not applicable to liquids or solids? What properties of liquids and solids indicate that their atoms or molecules are *not* far apart relative to their diameters?

13.2 In the model for gases referred to in Question 13.1, what is the justification for assuming collisions of the molecules with each other and with the walls of the container to be perfectly elastic? In answering this question, you should appeal to what we actually observe in the behavior of gases and what you would *expect* to observe if the collisions were *not* perfectly elastic. In other words, you must visualize, in the abstract, things that do not actually happen. (It is often the case in science that visualizing something that does *not* happen is just as important, and just as conducive to understanding, as knowing or visualizing what *does* happen.)

13.3 In the model referred to in Question 13.1, how do you interpret the macroscopic property of "density" of a gas in terms of molecular numbers? In visualizing the behavior of the continually moving atoms or molecules, how do you account for the *uniformity* of the density throughout any container? (When a system is uniform in all directions, we say it is "isotropic.")

13.4 In the model referred to in Question 13.1, how do you interpret, in terms of molecular behavior, the macroscopic pressure that any gas exerts on the walls of its container? In other words, what molecular behavior generates the pressure? How do you account for the *steadiness* of the pressure reading? In other words, why doesn't the needle of a pressure gauge keep jiggling randomly as individual molecules keep bumping the sensitive area of the gauge? How do you account for the uniformity of the pressure over the entire surface of the container? Suppose you had access to an extremely sensitive pressure gauge with an extremely small surface area. How would you expect the readings of this gauge to behave, especially as you decreased the amount of gas within the container? Explain your reasoning.

13.5 It is an observed fact that if we increase the temperature of a gas and keep the *volume constant*, the pressure of the gas increases. Suppose we now accept the idea that the gas consists of discrete particles or molecules. Argue that the observation we have described leads us to infer that increased temperature must be associated with increased molecular velocity. Explain why the constant volume restriction is a necessary part of the argument.

13.6 Balls of putty, each having a mass of 2.5 g, are projected in a steady stream against a wall as shown. The balls all have a velocity of 130 m/s perpendicular to the wall, and they stick to the wall on striking it. Over an interval of 5.0 s, 1600 balls strike a circular area of the wall having a diameter of 1.0 m.

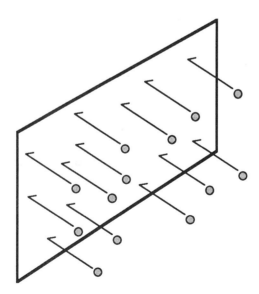

(a) Explaining your steps of reasoning, calculate the average pressure exerted on the wall by the rain of balls of putty.

(b) Compare your result in part (a) with the magnitude of normal atmospheric pressure: Is it large or small compared with atmospheric pressure? What significance do you see in the difference?

13.7 Suppose we start with a container of gas that has had a burner applied to one end so that the temperature of the gas at that end has been increased well above the temperature in the rest of the container. (Under such circumstances, we say that we have created a temperature *gradient* in the container.) We now remove the burner. Describe, in terms of molecular behavior and molecular kinetic energy, what happens as the gas returns to thermal equilibrium and the temperature gradient disappears. Explain the connection between the description you have just given on the microscopic level with what we previously called "transfer of heat" on the macroscopic level.

13.8 When we compress a gas by displacing a piston in a cylinder (as in a tire pump) against the opposing pressure of the gas, we must do work *on* the system to effect this compression. Describe in your own words what happens in the way of energy transformations in this process. There are several different cases to be considered:

(a) Consider the case in which the compression is performed fairly quickly and the cylinder is well insulated thermally from its surroundings. [In this process there is negligible heat transfer between our system (the gas, piston, and cylinder) and the surrounding air. Such a process is called "adiabatic."] In a *macroscopic* sense, what happens to the work we have done on the system? Into what other form or forms of energy has it been transformed? How do you describe and interpret the same process on the *microscopic* level? (Note: It is an observed fact that in adiabatic compression, the temperature of a *gas* always *increases*. In some rare instances involving other substances, e.g., water between 0 and 4 °C, temperature *decreases* on adiabatic compression.)

(b) If the cylinder is not thermally insulated and we compress the gas slowly, we can effect a compression in which the temperature of the gas remains essentially constant. Such a process is called "isothermal." Describe the energy transformations taking place in this process, first in macroscopic, and then in microscopic, terms.

13.9 Two glasses hold liquid water at room temperature: (1) one glass is open to the surrounding air, and (2) the other glass is tightly covered but has an air space above the water it contains. Imagine an initial condition in which we have just placed water in each container and there are, as yet, no water molecules in the air above the liquid.

(a) With sketches and verbal explanation, describe, in terms of the random motion of molecules of both water and air, what happens as time goes by after the initial condition specified above. What is the difference between cases 1 and 2? How do you account for the fact that all the liquid eventually disappears in case 1 while the liquid does not disappear in case 2? (As you have surely recognized, the technical name for the disappearance of the liquid under such circumstances is "evaporation.")

(b) In what sense is the term "equilibrium" relevant to the situation described in case 2? Is a condition of equilibrium one in which molecular motion and migration of molecules ceases? Is the term "equilibrium" relevant to case 1? Explain your answers in each instance.

(c) Compare any heat transfer that takes place between the air in the room and the liquid in cases 1 and 2: If any heat transfer does take place, what is the direction of the transfer? If no heat transfer takes place, say so explicitly. In either case, explain your reasoning.

13.10 A quantity of solid sugar is placed in a beaker containing liquid water. The sugar proceeds to dissolve in the water.

(a) With sketches and verbal explanation, describe what happens at the molecular level as the sugar dissolves, including a description of how the sugar molecules spread out (diffuse) through the water beyond the immediate vicinity of the solid sugar itself.

(b) Using a molecular level description similar to that given in (a), compare what happens when only a small amount of sugar is placed in the beaker, and all the solid disappears, with the situation that develops when a large chunk of sugar is placed in the beaker and the disappearance of the solid ceases at some point.

(c) In the light of the descriptions you have given , define the term "saturated solution." In what sense is the term "equilibrium" relevant to what is happening at saturation?

(d) It is well known that air dissolves in water. By means of sketches and verbal description similar to those you used in parts (a) – (c), describe what happens at the molecular level when an uncovered glass of water, initially free of dissolved air, is put out into the room. At what point does the amount of air dissolved in the water cease increasing?

Note to the student: In the following multiple-choice questions, circle the letters designating those statements you see to be <u>correct</u>. <u>Any number</u> of statements may be correct, <u>not</u> just one. You must consider each statement on its merits.

13.11 On the scale of relative atomic and molecular masses, nitrogen molecules (N_2) have a value of 28 while chlorine molecules (Cl_2) have a value of 71. With the two gases at the same temperature, the root mean square (rms) value of the velocity of the nitrogen molecules is

> (a) the same as the rms velocity of the chlorine molecules.
>
> (b) smaller than the rms velocity of the chlorine molecules by the factor 0.39.
>
> (c) is larger than the rms velocity of the chlorine molecules by the factor 2.5.
>
> (d) is larger than the rms velocity of the chlorine molecules by the factor 1.6.
>
> (e) None of the above.

13.12 If the velocity of every molecule in a fixed volume of gas were doubled,

> (a) neither the temperature nor the pressure would be altered because the volume is kept fixed.
>
> (b) both the temperature and the pressure of the gas would be doubled.
>
> (c) the temperature would be quadrupled and the pressure would be doubled.
>
> (d) both the temperature and the pressure would be quadrupled.
>
> (e) the specific heat of the gas would be doubled.
>
> (f) the average momentum of the molecules would be increased.
>
> (g) few containers would be strong enough to hold the gas.
>
> (h) None of the above.

CHAPTER 14

Modern Physics

14.1 The quantity 96,500 C emerges in measurements of amounts of material liberated in electrolysis experiments.

(a) In your own words, describe and illustrate the meaning of this quantity.

(b) Once we have accepted the premise that matter is constituted of discrete particles (atoms and molecules), what do the electrolysis measurements imply about how electrical charge is probably parceled out in the structure of matter? Is it likely to be continuous or discrete? Explain your inference carefully. (The inference you have articulated was made by Helmholtz and other scientists around the middle of the nineteenth century as the atomic-molecular model came to be accepted—a half-century before the experiments of Thomson and Millikan.)

14.2 Preceding Roentgen's discovery of X-rays in 1895, the process of electrolysis had been visualized as involving the migration, toward cathode and anode, respectively, of positively and negatively charged atoms (called "ions") of the elements forming the *compound* being decomposed in the electrolysis.

(a) Explain in your own words how the observations made in electrolysis experiments lead to and support the inference concerning ions. What does the notion (or "model") of ions suggest concerning how atoms might be held together in forming a molecule of a compound that makes a conducting solution on dissolving in water?

(b) With the discovery of X-rays, it was soon found that *all* gases could be made conducting by irradiating them with X-rays. Oppositely charged entities migrated to cathode and anode, even in the case of gaseous *elements* such as hydrogen, helium, neon, oxygen, and nitrogen. Thus the formation of "ions" was not limited to compounds as in electrolysis; *elements* also formed electrically charged particles.

What, if anything, does this discovery add to our view of the electrical structure of matter beyond the inferences already available from the electrolysis experiments?

14.3 Demonstration experiments are frequently performed with Crookes's tubes to illustrate properties of the cathode beam. If you have seen such demonstrations, consider the following tubes: (1) the tube containing a Maltese cross that can be swung up to intercept the cathode beam or swung down out of the way of the beam; (2) the tube with a small paddle wheel rolling on guides; (3) the tube that exhibits the effect of a magnet on the cathode beam.

(a) In each of the three demonstrations, describe what was actually *observed* to happen.

(b) In each of the three demonstrations, describe the *inferences* to be drawn about the properties of the cathode beam.

14.4 A droplet of oil or a small plastic bead, either one having a known density ρ (*mass* per unit volume) and known radius r, is injected into the space between capacitor plates as shown. The distance between the plates is denoted by d. After the air between the plates has been irradiated with X-rays, the bead becomes electrically charged (either positively or negatively).

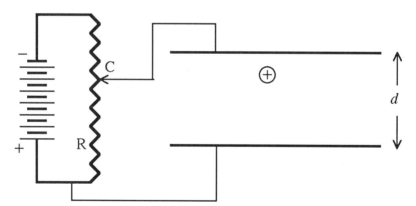

(a) How do you account for the bead becoming electrically charged? (It is a *fact* that it does.) Describe the processes you visualize taking place within the system, starting with the irradiation of the air and ending with excess charge on the bead.

(b) The potential difference ΔV between the capacitor plates can be varied by moving the contact C up or down along the resistor R, which is connected across the battery as shown in the diagram. Explain in your own words how this part of the system works: Why does the potential difference across the capacitor change as C is moved? Does the potential difference increase or decrease as C is moved up? What is the direction of the electrical field between the plates? Does the electrical field strength become larger or smaller in magnitude as contact C is moved up along R? Explain your reasoning.

(c) In the diagram, what would be the electrical effect on the charged particle if the contact C were located at the lowest point of the resistor R? Explain your reasoning. With the battery connected as shown, would it be possible to "balance" a negatively charged particle—i.e., keep it from falling under the influence of gravity? Why or why not? How would you connect the battery in order to balance a negatively charged particle?

(d) Draw a force diagram for the charged particle in the presence of a non zero electrical field and describe each force in words.

(e) Explaining your reasoning, obtain an algebraic expression for the *weight* of the bead or droplet in terms of the known quantities ρ, r, and g.

(f) Suppose that it turns out to be possible to "balance" the particle at some measured value of potential difference ΔV_o, i.e., the particle neither falls nor rises at the given electrical field strength. Show that the charge q present on the particle under these conditions can be calculated from the relation

$$q = \frac{4}{3}\pi r^3 \frac{\rho g\, d}{\Delta V_o}$$

14.5 Millikan, in his measurements of the amounts of electrical charge picked up by oil droplets in air that had been irradiated with X-rays, showed that the charge came in discrete "chunks" or "quanta" and that the size of the smallest chunk observed was 1.60×10^{-19} C.

Thomson had made measurements on positive ions formed in gases through which a cathode beam had been passed and reported that he had been able to observe both singly and doubly ionized species of every gas except atomic hydrogen. With atomic hydrogen he was able to observe only singly ionized atoms, i.e., ions carrying only one quantum of charge. It was also known in chemistry that hydrogen was never found to form a molecule in which two or more atoms of another element combined with only one atom of hydrogen; i.e., hydrogen was to be found only in combinations such as HX or H_nX but never as HX_n (where n denotes an integer number).

(a) Given the observed facts stated above and the additional fact that 1.01 g of hydrogen is liberated by passage of 96,500 C in electrolysis, calculate the number of atoms of hydrogen one would expect to have present in 1.01 g of the gas. Explain your reasoning, indicating the role played by each of the observations.

(b) Chemists have established that on a scale in which the commonly occurring carbon atom (C) has a relative mass of 12.0, hydrogen atoms (H) have a relative mass of 1.01, sodium atoms (Na) have a relative mass of 23.0, and chlorine atoms (Cl) have a relative mass of 35.5. Suppose we proceed to weigh out 1.01 g of H, 12.0 g of C, 23.0 g of Na, and 35.5 g of Cl. Explain what these very different masses of different materials must have in common. What significance do you now ascribe to the number calculated in part (a)?

(c) How many atoms altogether (Na and Cl combined) must be present in 58.5 g of common table salt, which has the molecular formula NaCl? How many *molecules* of the combined form NaCl must be present in the 58.5 g? Explain your reasoning in each instance.

(d) We know that ordinary water has the molecular formula H_2O and that the relative mass of the oxygen (O) atom on the scale cited above is 16.0. How many atoms altogether must there be in 18.02 g of water? How many *molecules* of the form H_2O? Explain your reasoning.

14.6 The density of solid crystalline sodium chloride (NaCl) is easily measured and is known to be 2.16 g/cm^3.

(a) Explaining your reasoning, calculate the volume occupied by 58.5 g of NaCl.

(b) Given the total number of atoms of Na and Cl present in the 58.5 g (using result obtained in Question 14.5), calculate the average spacing between centers of Na and Cl atoms in the solid material. Explain your reasoning with the help of a sketch of the geometry under consideration.

(c) The result you have obtained in part (b) yields a value for the approximate *size* (radius or diameter) of Na and Cl atoms. Explain why this is the case, being careful to indicate what role the observed fact that solids are very *incompressible* plays in this reasoning. Is the number you have obtained to be interpreted as a lower bound, an intermediate value, or an upper bound on the atomic sizes? Explain your reasoning.

(d) Using the well-known numerical value for the density of liquid water under ordinary conditions, calculate an approximate value for the radius or diameter of the water molecule (H_2O). Use appropriate sketches and explain all aspects of your reasoning, including whether you have obtained an upper or lower bound. Compare the values you have obtained in parts (c) and (d) and comment on the results.

14.7 Consider the following statement: "If, in the Thomson experiment, the two deflecting capacitor plates are brought closer together without changing the potential difference between them and without changing any of the properties of the cathode beam, the deflection of the spot on the screen would be observed to increase." Is this statement true or false? Explain your reasoning.

14.8 Consider the following statement: "Gravitational effect (curving toward the ground) of the beam in a cathode ray tube is not observed because the electrons in the beam have so small a mass that the gravitational effect is unobservably small." Is this statement true or false? Explain your reasoning. If you consider the statement false, how do you account for the fact that the gravitational effect is indeed unobservably small?

14.9 Consider the following statement: "As we decrease the pressure of gas in a discharge tube, we see a larger number of discrete lines in the emission spectrum because, at the lower gas density, the mean free path of the photons increases and more of them are able to get to the walls of the tube and escape from the gas." Is this statement true or false? Explain your reasoning. If you consider it false, how do you account for the fact that more discrete lines do indeed become visible?

14.10 A beam of positively charged particles passes between capacitor plates in a highly evacuated tube and strikes a fluorescent screen at the end of the tube as shown in the following diagram. When the capacitor plates are uncharged (switch S open), the beam makes a bright spot in the center of the screen at O. The particles in the beam all have the same charge and the same horizontal velocity v_x, but there are two populations of particles with two different masses m_A and m_B, with m_B greater than m_A.

(a) Sketch on the front view of the screen what you expect to see when the switch S is closed and the capacitor plates become charged. (The potential difference supplied by the battery is sufficient to deflect the beam but is not sufficient to drive the beam off the screen.) Be sure to mark the locations of m_A and m_B. Explain your reasoning.

(b) Suppose the beam were not homogeneous in velocity (i.e., both populations of particles have velocities ranging from some minimum value $v_{x\,min}$ to some maximum value $v_{x\,max}$.) Sketch what you would expect to see on the screen under these circumstances with switch S closed. Explain your reasoning.

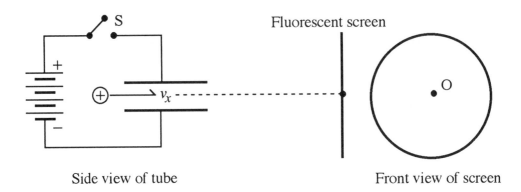

Side view of tube   Front view of screen

14.11 Consider an experiment like the Thomson experiment in which a beam of charged particles is passed between capacitor plates and strikes a fluorescent screen at the end of a tube, as shown in the following diagram. A magnetic field can be superposed in a direction perpendicular to the electrical field, i.e., into or out of the plane of the paper.

Suppose a beam of negatively charged particles is homogeneous in charge q and mass m but contains two distinct groups, one, consisting of half the particles, having initial horizontal velocity v_{1x}, and the other group having initial velocity v_{2x}, where v_{2x} is just twice as large as v_{1x}.

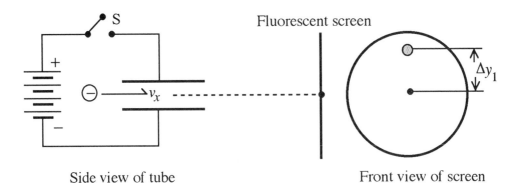

Side view of tube   Front view of screen

The circle at the right shows the screen. When electric and magnetic fields are zero and the beam is undeflected, the beam strikes the spot at the center. The spot a distance Δy_1 above the center is where the v_{1x} particles strike the screen when the switch S is closed and the beam is deflected by the E-field alone.

(a) Add to the picture whatever is necessary to show what the screen will look like when both groups of particles are present. That is, will there be a second spot or will there be a smearing out of the screen pattern? If there is a second spot, put it on the screen, positioning it correctly in relation to the v_{1x} spot. (The position you show must be correct relative to the scale defined by the length Δy_1 on the diagram, and some ratio reasoning may therefore be required.) Explain your reasoning.

(b) In the manner exploited by Thomson, a magnetic field is now introduced in a direction perpendicular to the plane of the paper, and its strength is adjusted (by varying current in the coils producing it) until the v_{1x} particles are returned to the center of the screen. Show what the screen will now look like. Where, that is, will the remaining particles strike the screen? Explain your reasoning.

14.12 In an experiment measuring the photoelectric properties of a certain metal, it is found that the threshold for photoemission is at a certain frequency of violet light v_o.

(a) Suppose the incident light is now changed to a frequency in the ultraviolet region without change in the intensity. What, if anything, will happen to the observed stopping potential and the photocurrent? Explain your reasoning.

(b) Suppose the incident light is now changed to a frequency in the blue region without change in the intensity. What, if anything, will happen to the observed stopping potential and the photocurrent? Explain your reasoning.

(c) Suppose the frequency of the incident light is increased to a value $1.43\ v_o$. What will be the maximum kinetic energy of emitted electrons? Explain your reasoning.

14.13 The radius of an atom is of the order of 20,000 times the radius of an atomic nucleus. Let us take the order of magnitude of the density of liquids and solids to be of the order of 2 g/cm^3. Suppose that atoms were stripped of their cloud of electrons, and the bare nuclei were packed as closely as atoms are in liquids and solids.

(a) Using ratio reasoning, calculate the order of magnitude of the density of matter consisting of closely packed atomic nuclei. (To visualize the significance of your result, recalculate it in units that might be more directly familiar. Try tons per cubic inch, for example.) There is reason to believe that certain stars consist of matter approaching, and even exceeding, this fantastic density.

14.14 After publication of Einstein's suggestion of the photon model, various individuals tried to preserve the older wave picture by proposing modifications of classical theory. Among these efforts was one by J. J. Thomson [*Proc. Camb. Phil. Soc.* **XIV**, 417 (1907)]. Thomson suggested that electromagnetic energy might be unevenly distributed over the wave front, with regions of maximum energy relatively widely separated by areas of low or zero disturbance. This hypothesis led to the suggestion that at extremely low light intensities, when only a few maximum energy regions would be present on any given wave front, ordinary diffraction patterns formed by slits or shadows of small obstacles might be modified in some observable way, perhaps by fuzzing out of the pattern

An experiment to test this hypothesis was carried out by G. I. Taylor [*Proc. Camb. Phil. Soc.*, **XVI**, 114 (1909)]. Taylor reports that:

"Photographs were taken of the shadow of a needle, the source of light being a narrow slit placed in front of a gas flame. The intensity of the light was reduced by means of smoked glass screens. . . . The longest time [of exposure with very weak light] was 2000 hours or about 3 months. [Legend has it that Taylor, being an avid sailor, went off sailing during this period.] In no case was there any diminution in the sharpness of the pattern. . . . The amount of energy falling on one square cm of the plate is 5×10^{-6} erg/sec, and the amount of energy per cubic cm of this radiation is 1.6×10^{-16} erg.''

(a) Show that from the point of view of the photon model, the given energy flux of 5×10^{-6} erg/(s) (cm^2) corresponds to about 10^6 photons of visible light per second per square centimeter (taking the average energy per photon to be about 2 or 3 EV).

(b) Show that this flux of photons implies that the average distance of separation between individual photons must have been of the order of 300 m.

With an apparatus of the order of 1 m in length, it is extremely unlikely that more than one photon would have been present in the system at the same time! With modern counting equipment and photomultiplier tube detectors, Taylor's experiment is readily repeated without waiting 3 months for a sufficient photographic exposure. All such experiments confirm Taylor's original results. For example, a two-slit interference pattern formed by an intense beam of light is identical with one formed by light so weak that *only one photon* is likely to be passing through the slit system at any given instant. Closing either one of the two slits eliminates the interference pattern.

(c) What do you infer from the fact that an interference pattern is still formed under conditions of extremely weak illumination? How can two slits be effective in producing an interference pattern when only one photon arrives at a time? [Do not expect to give or receive a simple, pat answer to this question; the problem being posed lies at the heart of modern quantum physics, but there is every reason for you to begin to think, and wonder, and speculate about it.]

Note to the student: In the following multiple-choice questions, circle the letters marking those statements that are true or <u>correct</u>. <u>Any</u> <u>number</u> of statements may be correct in a given question, <u>not</u> necessarily just one. It is necessary to examine each statement on its merits.

14.15 The measurements of alpha particle scattering made by Geiger and Marsden and interpreted by Rutherford showed for the first time that

(a) protons are more massive than electrons.

(b) atoms contain a very dense concentration of positive charge.

(c) neutrons must be present in atomic nuclei.

(d) alpha particles must be positively charged.

(e) the region of dense concentration of positive charge must be very much smaller than the size of the atom.

(f) oscillating electrically charged particles must radiate electromagnetic waves.

(g) electrons circulate around the nucleus in circular or elliptical orbits.

(h) alpha particles are scattered by target atoms through large as well as small angles.

(i) None of the above.

14.16 [In this question you must be careful to discriminate between what is *observed* and what is *inferred* from observations.] In the photoelectric effect, light incident on a metal surface causes electrons to be ejected from within the metal. If the <u>intensity</u> of an incident beam of monochromatic light is <u>decreased,</u> it is an OBSERVED FACT that:

(a) the rate of ejection of electrons decreases.

(b) the photoelectric current ceases below a certain threshold of minimum intensity of the incident light.

(c) the maximum kinetic energy, possessed by the ejected electrons at the point at which they leave the metal surface, decreases.

(d) the observed photoelectric current decreases.

(e) the potential difference that just brings the photoelectric current to zero remains unchanged.

(f) electrons are bound within the metal by a certain minimum amount of energy referred to as the "work function."

(g) None of the above.

14.17 Millikan, in his famous oil drop experiment,

(a) measured the charge carried by the cathode ray particles (electrons) detected in discharge tubes such as those used in the Thomson experiment.

(b) demonstrated that both positive and negative electrical charges are quantized, i.e., come in discrete packages of identical finite size.

(c) made it possible to calculate a reasonably precise value of Avogadro's Number.

(d) demonstrated that atoms have very tiny positively charged nuclei with electrons located relatively far from the nucleus.

(e) took into account the frictional effect between the droplet and the surrounding air in those observations in which terminal velocities of the droplets were being observed.

(f) neglected the effect of gravity compared with the electrical force on the charged droplet.

(g) had to find the smallest common multiple among the charge quantities he measured because the droplets, on different occasions, carried varying numbers of "packages" of charge.

(h) None of the above.

14.18 In the simplest Bohr model of the hydrogen atom, with the electron visualized as executing purely circular orbits around the proton:

(a) The resulting formula for the allowed orbits shows that the radii of these orbits increase in equal steps of length with increasing value of the quantum number n.

(b) Neglect of the gravitational force between the electron and proton is one of the basic faults of the model.

(c) A "stationary state" is one in which the electron has zero velocity.

(d) The electron orbit should be surrounded by a magnetic field similar to that surrounding a current-carrying loop of wire, and each hydrogen atom should therefore behave like a minute magnet with north and south poles.

(e) In the expression $E(r) = -ke^2/2r$ for the total energy associated with an orbit of radius r, the 2 in the denominator arises in the calculation of the potential energy associated with the Coulomb interaction in the electron–proton system.

(f) A higher energy photon is released when the electron jumps from orbit $n = 2$ to orbit $n = 1$ than is released when the electron jumps from orbit $n = 100$ to orbit $n = 2$.

(g) The kinetic energy of the electron is larger in orbit $n = 4$ than in orbit $n = 2$.

(h) The orbital kinetic energy of the electron approaches zero as the radius of the orbit increases without limit.

(i) The negative sign associated with the total energy of the system in any allowed orbit stems from the fact that the potential energy decreases more than the kinetic energy increases as the radius decreases from very large values.

(j) In the "ground state" of the atom, the radius of the electron orbit is essentially zero.

14.19 According to the Bohr model of the hydrogen atom:

(a) a photon of energy $hv = 2\pi^2 m(ke^2)^2/h^2$ would be capable of ionizing a hydrogen atom.

(b) the reddest line in the Balmer spectrum involves an electron transition from orbit 3 to orbit 2.

(c) transitions are possible between <u>any</u> <u>two</u> allowed orbits.

(d) if an incident photon, absorbed by an electron in the ground state, has just the right energy to kick the electron from orbit 1 to orbit 5, as many as four different photons might be emitted as the electron cascades back to the ground state.

(e) the electron would have nonzero angular momentum in all allowed states including the ground state.

(f) None of the above.

14.20 How are some of the X-rays produced in a cathode ray tube in which the electron beam strikes a metallic target?

(a) The electrons, when accelerated through a large potential difference, are converted into X-rays on striking the target.

(b) Photons are accelerated by the high potential difference until their energy lies in the X-ray region.

(c) The electrons emit X-rays as they are accelerated initially.

(d) The electrons emit electromagnetic radiation when they are slowed down on striking the target.

(e) X-ray photons, present in the target, are released when electrons strike the target.

(f) X-ray photons are liberated from the target by the photoelectric effect.

(g) Electrons annihilate positrons in the target, producing electromagnetic radiation.

(h) None of the above.

14.21 In the Bohr model of the hydrogen atom, an appropriate force diagram (or diagrams) for the two interacting particles would be:

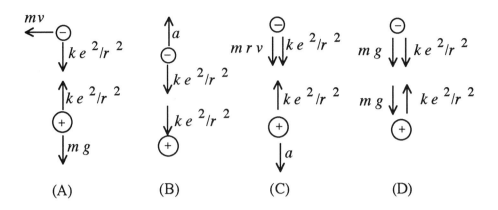

(A)   (B)   (C)   (D)

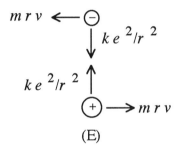

(E)

(F) None of the preceding

14.22 A beam of X-rays passes near a charged electroscope. The leaves of the electroscope will collapse if the electroscope is

(a) confined in a highly evacuated chamber.

(b) surrounded by a gas.

(c) charged positively, but not if it is charged negatively.

(d) charged negatively, but not if it is charged positively.

(e) not subject to a magnetic field.

(f) None of the above.

14.23 It is an observed fact that the emission spectrum of an atom (e.g. sodium) has many more bright lines than the absorption spectrum has dark lines. This difference between emission and absorption spectra is explained by the fact that

(a) fewer atoms are absorbing light when absorption spectra are formed than are emitting light when emission spectra are formed.

(b) in absorption, electrons are not elevated to energy levels as high as those from which they start in the case of emission.

(c) in absorption electrons are all elevated to higher energy levels from the ground state while in emission they cascade back to the ground state through intermediate states.

(d) the ionization potential of the atoms is a variable quantity.

(e) the magnet moments of the atoms are different in different energy levels.

(f) None of the above.

CHAPTER 15

Mixed Areas of Subject Matter

15.1 The diagram shows a system in which a glider with mass m_G, starting from rest, slides down an air track inclined at an angle θ to the horizontal. It descends along a length L of the track while dropping through a vertical height H. Simultaneously, a ball with mass m_B also starting from rest, drops vertically through the same height H. We shall compare these two motions in two different ways and analyze their similarities and differences. (Frictional effects are to be taken as negligible in the analysis, which is to be carried out entirely algebraically.)

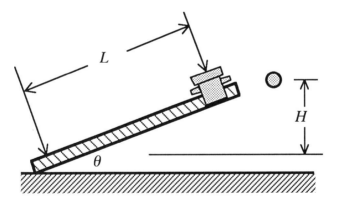

(a) What is the linear acceleration of the glider? How does it compare with that of the ball? Explain your reasoning.

(b) Use the kinematic equations to obtain expressions for the velocities v_G and v_B of the glider and the ball after each has dropped through the height H, as well as expressions for the time intervals Δt_G and Δt_B for the descent of each object to this final level.

(c) Use a conservation of energy argument to obtain expressions for both v_G and v_B.

(d) Analyze and interpret the results you have obtained in parts (b) and (c). How do the two velocities compare? Which approach is the more powerful for obtaining the velocity information? How do the two time intervals compare? That is, which is longer? How do you explain the difference? Analyze the time difference algebraically: How does it vary as the angle of inclination of the track is changed without changing H? Show that

$$\Delta t_G - \Delta t_B = \sqrt{\frac{2H}{g}}\, \cot\theta$$

(e) Could you have obtained the time difference information directly from the energy argument alone? Why or why not?

15.2 A plane is inclined at an angle of 22.0° to the horizontal as shown. A block with a mass of 12.0 kg, sliding down the plane under the influence of friction, passes position A, with its center of mass at a height of 4.0 m above the ground, and with a velocity of 3.52 m/s. It slows down under the influence of the substantial frictional force, coming to a stop at position B, with its center of mass 0.5 m above the ground.

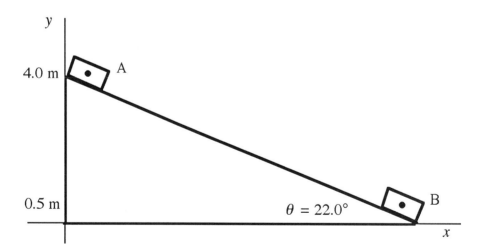

(a) Draw force diagrams of (1) the block at some intermediate position between A and B and (2) of the region of the plane in contact with the block at that same position. Describe each force in words.

(b) Explaining your reasoning, calculate the increase in *thermal* internal energy of the system consisting of the block and the plane, assuming that there is negligible heat transfer to the surrounding air as the temperatures of the block and plane increase.

(c) Between the initial and final conditions, what is the total change in internal energy of the entire system consisting of the earth, the block, and the plane? Explain your reasoning.

15.3 A ball is projected in a trajectory such as that sketched. Exactly at the top of its flight it undergoes a perfectly inelastic, head-on collision with an identical ball that is suspended on a very weak thread, which breaks on the collision.

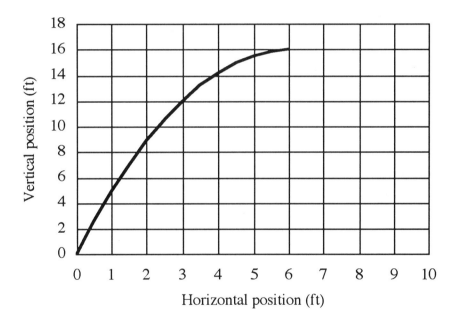

(a) Sketch on the diagram the continuing trajectory of the two-ball combination after the top of the flight to the point at which it lands on the ground. Explain how you drew the trajectory.

(b) Defining your system clearly and carefully, describe all the momentum and energy changes that take place as the first ball is on the way up, as the two balls collide, and as the two balls continue to the ground.

Note to the instructor: Question 15.4 is intended to help students understand the concept of "components of a displacement vector" by connecting this idea to the displacement of the shadow of an object along a wall.

    15.4 Consider the situation illustrated in the following diagram: two walls, A and B, are shown perpendicular to each other. A small ball is located initially at point P. Two fairly distant search lights (one directing a parallel beam directly at wall A, and the other a parallel beam directly at wall B) cast sharp shadows of the ball on each wall.

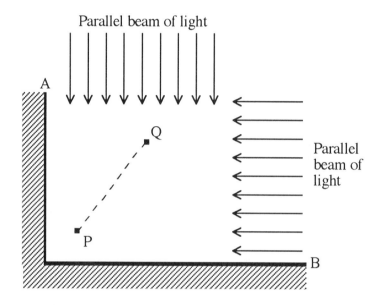

(a) Denoting the shadow points by the symbols P_A and P_B, respectively, mark the location of the shadow of the ball on each wall.

    The ball is now moved from P to Q, a distance of 3.76 m, along the dashed line. This line makes an angle of 52.5° with wall B and lies in the plane that is represented in the diagram.

(b) Denoting the shadow points by the symbols Q_A and Q_B, respectively, mark the location of the shadow of the ball on each wall.

(c) Calculate the distances $P_A Q_A$ and $P_B Q_B$, i.e., the distances between the shadows on each wall. Explain your reasoning.

(d) Return to the concepts of "displacement vector" and "rectangular components of a displacement vector" that you studied in mechanics. Explain the connections between the calculations you made in part (c) and these vector concepts. Explain in your own words why rectangular components of *velocity* vectors can be similarly calculated.

Note to the instructor: In general, students have not been led to look at tables of data to discern significant relations, insights, or generalizations that are frequently contained in the data. They tend to view tables as useful only for finding numbers to put into end-of-chapter problems. Following is an example of what might be done with a fairly mundane table, that of densities of various materials.

15.5 This table gives the densities of various materials at room temperature (20 °C) and atmospheric pressure.

Table of Densities (g/cm^3)					
Solids		Liquids		Gases	
Material	Density	Material	Density	Material	Density
Aluminum	2.70	Carbon tetrachloride	1.59	Air	1.20×10^{-3}
Brass	8.5	Ethyl alcohol	0.791	Ammonia	0.77×10^{-3}
Copper	8.96	Gasoline	0.66–0.69	Carbon monoxide	1.16×10^{-3}
Gold	19.3	Mercury	13.6	Carbon dioxide	1.84×10^{-3}
Iron	7.87	Methyl alcohol	0.810	Chlorine	3.00×10^{-3}
Lead	11.4	Milk	1.028–1.035	Helium	0.165×10^{-3}
Platinum	21.4	Oils	0.92–0.97	Hydrogen	0.084×10^{-3}
Silver	10.5	Seawater	1.025	Methane	0.67×10^{-3}
Sodium	0.97	Sulfuric acid	1.84	Oxygen	1.33×10^{-3}
Tin	5.75	Water	1.00	Nitrogen	1.16×10^{-3}
Uranium	19.0	Liquid air (at −193 °C)	0.87	Sulfur dioxide	2.73×10^{-3}
Zinc	7.13			Air at 10 km altitude *	0.37×10^{-3}
Ice	0.917				
Glass	2.4–2.8			Air at 20 km altitude *	0.09×10^{-3}
Plastics	0.90–1.1				
Salt	2.18				
Sugar	1.59				
Stone	2.4–3.1			* in situ values, *not* at 20 °C and 1 atm.	
Wood Balsa	0.11–0.14				
Oak	0.60–0.90				

(a) Examine the table from the standpoint of comparing the general categories of solids, liquids, and gases: Is there a general trend of increase or decrease in density as one shifts from one category to another? Which way? To what extent, if any, do the categories overlap, or not overlap? What, if anything, seems special about gases? What is the order of magnitude (round number) of the ratio of the density of gases to that of liquids?

(b) What are the most dense materials listed? The least dense? Are you surprised by the position of lead in the sequence? Suppose you put a piece of lead in a beaker of mercury. How would the lead behave? How many of the listed materials would sink in mercury? Are there any metals that would float on water (if prevented from reacting explosively with it)?

(c) Suppose you wished to concoct a liquid that is not water but has a density very nearly equal to that of water. On the basis of the table, what liquids might you try to mix together to achieve this? If you were actually to undertake such a task, what had you better find out about various properties and behaviors of the liquids you wish to mix?

(d) Air is actually a mixture of the gases oxygen and nitrogen (roughly 20% oxygen). Suppose the gases separated in the atmosphere instead of remaining mixed. Which gas would end up "floating" on top?

(e) Scientists of the late eighteenth and early nineteenth centuries found it hard to see why the gases did not separate, with the nitrogen floating on top. The fact that the gases remained mixed puzzled them. From your present vantage point, how would you explain the situation to one of those scientists and make your explanation convincing?

Note that the density of gases is very, very much less than that of liquids or solids. Note also that gases are highly compressible, very much more so than liquids or solids. [You are aware of this although you may never have articulated it explicitly. "Highly compressible" means that it is easy to decrease the *volume* of a gas very substantially by squeezing it (as in a bicycle pump), but it is extremely difficult to change the volume of liquids or solids by squeezing. The volume of the latter *can* be decreased somewhat, but enormous pressures are required to produce significant changes. Liquids and solids are therefore described as "relatively incompressible".]

(f) If we accept the view that material substances consist of discrete particles (atoms and molecules), what do the observed densities and facts about compressibility suggest about the relative spacing of particles in gases on the one hand and liquids and solids on the other? Explain your reasoning. In the light of the round number density ratio you noted in part (a), what is the order of magnitude of the ratio of average spacing between particles in gases at room temperature and atmospheric pressure on the one hand and the average spacing in liquids or solids on the other? Explain your reasoning.

Note to the instructor: Question 15.6 connects thinking in electrical and mechanical contexts.

15.6 The direction of the electrical field everywhere along the surface of a charged metallic conductor must be normal to the surface after the charge has settled down into an equilibrium distribution.

(a) Why is this the case? What would happen to any excess charge in a region of the surface where the field direction was *not* normal to the surface? (Draw simple diagrams to illustrate your argument.)

(b)Must the electrical field direction be everywhere normal to the surface of a charged nonconductor? Why or why not?

(c) Compare your argument about the electrical situation in the charged metallic conductor with the mechanical situation shown in figure (a) in which a block is held at rest on a sloping air track (negligible friction) by means of the balancing force exerted by string A parallel to the track: What must be the direction of the force exerted by the air track on the block under this equilibrium condition? Why is this force given the name "normal force"? What would be the effect on the block if this force were not acting in a direction perpendicular to the track but had a component one way or the other along the track?

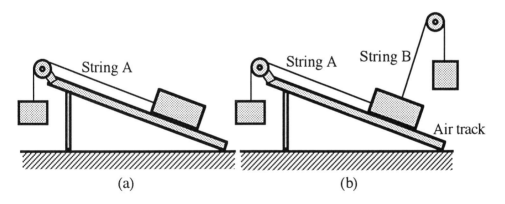

(a)   (b)

Suppose the effect of the track in figure (a) is replaced by the pull of string B as shown in figure (b), and the block is just barely lifted off the track.

(d) What is the equilibrium orientation of string B? Suppose you displace the block slightly up or down the plane from the equilibrium position so that string B is no longer normal to the track. What will happen to the block after you let it go from the displaced position?

(e) Explain the analogy between this mechanical situation and that of a charged region in the metallic conductor where the electrical field is not initially normal to the surface?

Note to the instructor: Question 15.7 has several purposes. One is to spiral back to the centripetal force ideas as soon as Coulomb's law is available. Another is to confront students with a situation in which the applied force may be either larger or smaller than that necessary to impart the indicated centripetal acceleration, and the motion may not then be circular. (Students rarely encounter such situations, and they develop the habit of substituting indiscriminately into the formula applicable to circular motion as though it were universally valid.) The third purpose is to present a situation in which the student is not explicitly told what to calculate and must make the decision independently. This is a very rarely available opportunity.

   15.7 We are looking down on two small frictionless pucks A and B located on a level air table. Firmly fastened to each puck is a small uniformly charged sphere, as indicated. Puck A is firmly fastened to the table and does not move. Its sphere carries a charge of $+5.0 \times 10^{-7}$ C. The sphere on puck B carries a charge of -8.0×10^{-7} C, and the mass of B (with its charged sphere) is 125 g. At a given instant, Puck B has an instantaneous velocity of 0.45 m/s in a direction perpendicular to the radial line between A and B. The radial separation of the pucks at this instant is 6.3 cm.

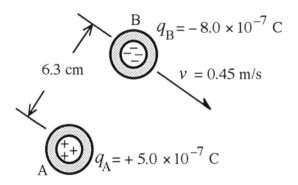

(a) We raise the following question: Will puck B follow a circular path around A? If not, will it tend to increase or decrease its radial distance from A? Note that you are not being told exactly what to calculate; you must make the decision yourself. That is one of the main points of the problem. Show all your numerical calculations, explaining why you make each one and explaining the inferences to be drawn from your calculations. Handle units carefully. Be sure to draw force diagrams for both pucks A and B.

15.8 A heating element with a resistance of 25 Ω is designed to be connected to a 120 V source of electric current.

(a) Explaining your reasoning by giving relevant definitions, calculate the wattage rating of the heating element.

The heating element is now immersed in 1000 g of water in a well-insulated brass calorimeter at an initial temperature of 20 °C and is allowed to run for 15.0 min, after which it is removed from the water. The brass container has a mass of 250 g and a heat capacity of 0.10 cal/(g)(C°).

(b) Explaining your reasoning as you go along, predict the final condition in the calorimeter. That is, what will be the final temperature? Will any of the water have been boiled away? If so, how much?

15.9 In our investigation of the magnitude of atomic and molecular dimensions, we found that this was of the order of 2 or 3 $\overset{o}{A}$. In measurements of the wavelength of visible light, we have found such wavelengths to be of the order of 5000 $\overset{o}{A}$. Recall what happens to a wave train that passes by an object much smaller than its wavelength.

(a) In the light of the insights and experiences listed above, discuss the possibility of our actually "seeing" an atom or molecule by illuminating it with visible light. Appeal, for example, to what you might have observed when water waves or ripples are incident on large or small obstacles. Explain your reasoning carefully.

15.10 Through observing objects or systems "doing things" to each other, we become aware of interactions in which accelerations can be imparted. Through Newton's first law, we associate such accelerations with the action of forces, whether or not we observe direct physical contact between the interacting objects. We have, so far, in macroscopic phenomena, separated observed interactions not involving direct physical contact into three classes having the names "gravitational," "electric," and "magnetic." At this point let us review these effects and see them in perspective.

(a) By listing (1) conditions under which each of these interactions is observed, (2) effects and changes we can produce through our own manipulation of the interacting objects, and (3) specific *differences* we discern among the interactions, show why we are justified in distinguishing three different classes of phenomena rather than lumping them into one. Be specific in recognizing similarities as well as differences. Your listing should be highly "operational"; in other words, you should describe specific things we can do and the consequences of doing them. [Some hints as to relevant aspects: What specific substances are involved? How do different substances differ in their behavior under relevant treatment? Are relevant properties transferable (or not transferable) from one object to another on contact? What changes can be effected by touching or otherwise manipulating the interacting objects? What evidence is there for presence or absence of "induction" effects? Are there differences between any of the observed induction effects? And so forth.]

15.11 A formula, standing by itself, is nothing but a collection of letters or symbols. A formula is given meaning only by the *text* that goes with it. The text contains (1) a description of the meaning of the symbols, (2) the physical situations to which the formula applies (or does not apply), and (3) the respective roles of definition, empirical result of experiment, or broader "law of nature" that enter into or are expressed in the formula.

(a) Write a paragraph about each of the following formulas, describing its meaning in terms of the characteristics cited above. Take into account the fact that *some* formulas are *true* because *all* the quantities that enter into them are defined and no knowledge of behavior in nature is therefore necessary to make them true. Such formulas may apply, however, to *some* natural phenomena.

(1) $v = v_0 + at$
(2) $F_{net} = ma$
(3) $f_{max} = \mu N$
(4) $F_{elect} = kq_1 q_2 / r^2$
(5) $pV = \text{constant}$

(6) $s = s_0 + v_0 t + (1/2)at^2$
(7) $F = kx$
(8) $F_{cent} = mv^2/r$
(9) $F_{grav} = Gm_1 m_2 / r^2$
(10) $\Delta V = IR$

15.12 Consider the following classes of uniform circular motion: (1) a bob on a string in a circle lying in a horizontal plane or a frictionless puck tied to a fixed point on a level air table and (2) a satellite in circular orbit around the earth or a negatively charged particle in circular orbit around a fixed positively charged particle.

By appealing to appropriate physical laws and the resulting equations, show that in case 1, any angular velocity is possible at a given radius up to the point at which the string breaks, but that in case 2 only one angular velocity is possible at a given radius. Explain this profound difference between the two cases.

15.13 Suppose we have a frictionless puck of mass m attached to a string on a level air table, as shown. The puck revolves in a circle around point O where there is a hole in the table. The string can be pulled down through the hole at O, decreasing the radius r of the circle in which the puck moves. The tangential velocity of the puck at any radius r is denoted by v_T. Suppose we pull down on the string, decreasing r very slowly, and imparting a very small radially inward velocity v_r to the puck in addition to its tangential velocity v_T.

We know that the puck speeds up as we shorten the radius, increasing its tangential and angular velocities in a manner analogous to the increase in angular velocity of a skater pulling in his or her arms. But note that there is something paradoxical about this situation: there is no external torque applied to the system of puck and string since the force is directed along the string and has no component perpendicular to it. How does the puck manage to speed up in the absence of external torque? Is there something wrong with our dynamical theory? In this problem, we shall analyze, in detail, the speeding up of the puck.

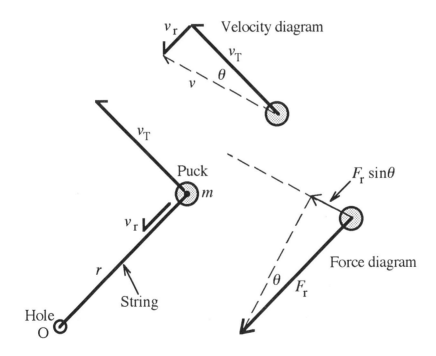

In the velocity diagram for the puck, the resultant of velocities v_T and v_r is a velocity v in a direction lying at a small angle θ below the tangential direction. We shall call θ the "angle of descent." (The idea is to keep v_r very much smaller than v_T; the scale of the diagram is greatly exaggerated to make the effect of v_r clearly visible.)

The force diagram for the puck is also shown. Note that the centripetal force F_r has a nonzero component in the direction of descent, i.e., along the resultant velocity v. It is this component of the force (in the presence of the small radial velocity v_r) that accelerates the puck. Now we know, qualitatively, how the puck gets to be accelerated in the absence of an external torque, but we do not have a quantitative analysis. What we want is an equation relating v_T and r directly so that we can see what happens to the tangential velocity as we pull the puck in or let it out. Let us analyze the situation algebraically, starting with the acceleration imparted when v_r is not zero.

(a) Let us denote the acceleration in the direction of descent by a_D. Noting how the accelerating force is related to the centripetal force on the force diagram, argue that

$$a_D = \frac{v_T^2}{r} \sin \theta \tag{1}$$

(b) We can replace $\sin \theta$ by its connection with the velocities v_r and v on the velocity diagram, but we must now be very careful about algebraic signs because v_r can be either positive (outwardly directed) or negative (inwardly directed). From the velocity diagram it follows that we can write

$$a_D = -\frac{v_T^2}{r} \frac{v_r}{v} \tag{2}$$

where we have introduced the minus sign to keep a_D positive when v_r is inward and negative when v_r is outward.

(c) If v_r is very small (i.e., if we pull the puck in or let it out very slowly), the resultant velocity v is very nearly equal to v_T. Argue that we can alter eq. (2) to

$$a_D \cong -\frac{v_T v_r}{r} \tag{3}$$

(d) Since our objective is to obtain a relation between v_T and r, we want to get rid of a_D in terms of these quantities. For this purpose we can make powerful use of the chain rule of differentiation:

$$a_D = \frac{dv}{dt} = \frac{dv}{dr} \frac{dr}{dt} \tag{4}$$

Show that eqs. (3) and (4), combined with the fact that v is very nearly equal to v_T, yield the simple differential equation:

$$\frac{dv_T}{dr} \cong -\frac{v_T}{r} \tag{5}$$

(e) Solve the foregoing differential equation [eq. 5] to show that the quantity rv_T is constant and this, in turn, implies that $r^2\omega$ is constant (where ω is the angular velocity of the puck). In other words

$$r^2\omega = r_o^2\omega_o \tag{6}$$

where r_o and ω_o denote, respectively, any arbitrary initial starting radius and angular velocity. Interpret eq. (6): What happens to the angular velocity as the puck is pulled inward? As it is let outward?

(f) Argue that the analysis we have carried out shows that the angular momentum of the system is conserved, as it should be in the absence of external torque, but that the puck is nevertheless accelerated tangentially as it is slowly moved radially.

(g) Carry out an integration to obtain an expression for the work done on the system in pulling the string inward (or letting it out), and show that this amount of work turns out to be equal (as one might expect) to the change in kinetic energy of the puck. (Be very careful about algebraic signs.)

(h) Why has our analysis been confined to pulling the puck in or letting it out *very slowly*? How does the puck behave if we pull it in rapidly? (Don't try to analyze the latter situation algebraically; just try to visualize it physically—or, better yet, try it out experimentally.)

15.14 The following table summarizes typical values of energy associated with various particles, photons, and physical changes. You can begin to understand and appreciate the pattern and scale of energy transformations in the microscopic domain by viewing and assimilating the various orders of magnitude exhibited in this table. The simple calculations and questions in parts (a) through (f) will help you begin to familiarize yourself with the table and its significance.

(a) Fill in the blanks in the right-most column (column 4) by calculating the corresponding value of wavelength λ for each instance in which a value is not already entered.

			Wavelength λ of photon of
	Substance, particle, or	Energy in electron	same energy
Type of energy	photon	volts (ev)	(Å)
Average total energy of vibration an atomic particle along one coordinate axis in a solid material at room temperature	—	0.02 (approx.)	
Average total translational kinetic energy of a gas molecule at room temperature	—	0.04 (approx.)	
Average energy released per atom in a violent chemical reaction (e.g., the explosion of TNT)	—	0.5 (approx.)	
First ionization potential (energy necessary to separate one electron from an isolated, neutral atom)	K Na Ag Pt Au H He	4.3 5.1 7.5 8.9 9.2 13.6 24.5	 505
Photoelectric work function (minimum energy necessary to eject electron from surface of a metal)	K Na Ag Au Pt	3 3 4.7 4.8 6.3	
Photons of various types of electromagnetic radiation	Far infrared Sodium D lines Ultraviolet X-rays	0.02 2.1 6 Tens of thousands	 5890
Radioactive emanations	α particles from U and Ra β particles from many isotopes γ rays from many isotopes	4–6 MeV 0.1–2 MeV 0.01–1 MeV	

Comparison of Energies Associated with Various Microscopic Particles and Phenomena

(b) We associate the increase in temperature of any material with an increase in the kinetic (or "thermal") energy of the atoms or molecules of the material. Note the order of magnitude of such energies as indicated in the first two entries of column 3. Note also that your calculations in column 4 show that photons of such energies have wavelengths in the far infrared. Why do you suppose that infrared radiation is frequently referred to as "thermal radiation"?

(c) The average translational kinetic energy E of gas molecules is given as a function of absolute temperature T by the following equation: $E = 1.29 \times 10^{-4} T$, where E is in electron-volts. At what temperatures would the average kinetic energy of gas molecules be in the range of energies of photons of visible light? (Answer: Temperatures of the order of 15,000 K.) Note that the temperature of the surface of the sun is about 6000 K.

(d) Note the enormous difference between the order of magnitude of energy per atom in a violent chemical reaction and that associated with radioactive disintegration. What implications do you see in this difference?

(e) How do you account for the fact that X-rays and radioactive emanations cause ionization of gases while visible light does not? What implications does this have with respect to the effects one might expect X-rays and radioactive emanations to have on complex molecules such as those that, for example, make up biological materials?

(f) The work function for ejecting electrons from a metal is lower than the ionization energy of an isolated atom of the same substance. What implications do you see in this difference?

You would do well to note and remember the various orders of magnitude illustrated in this question. This will help you think qualitatively and powerfully about a vast range of physical phenomena occurring in the world around us.

15.15 You are informed that in an empty compartment, to which you have access with instruments, there exists either an E-field or a B-field but not both. Describe several experiments or observations you might perform in the empty space to determine which kind of field is present.

15.16 Consider the following thought experiment: an electron and a proton (hydrogen ion), released from rest at opposite plates of a capacitor in a highly evacuated space, are accelerated across the capacitor and exit through holes in the plates as shown in the following diagram. The potential difference between the plates is 150 V. The mass of a proton is very much larger than the mass of an electron.

(a) Explaining your reasoning, compare the kinetic energies of the two particles as they exit through their respective openings. ("Comparing" means establishing whether one quantity is greater, equal to, or smaller than the other.) It is not necessary to calculate numerical values of the kinetic energies.

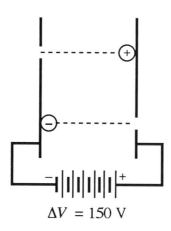

$$\Delta V = 150 \text{ V}$$

(b) Explaining your reasoning, compare the momenta of the two particles as they exit through their respective openings. It is not necessary to calculate numerical values of the momenta.

Note to the student: In Question 15.17 circle the letters marking <u>all</u> those statements that are <u>correct</u>. <u>Any number</u> of statements may be correct, and each one must be examined carefully on its merits. Do not simply abandon the question when you have found one correct statement.

15.17 A projectile fired from point A at ground level has initial velocity v_0 inclined at angle θ to the horizontal, as shown in the following diagram. If the air resistance were negligible, the projectile would follow a parabolic trajectory through point P, returning to the ground at point B. The range of the trajectory is denoted by Δx_m. The following statements are to be taken as applying to this idealized situation.

(a) The magnitude of the instantaneous velocity at point P (top of the flight) is given by $v_0 \cos \theta$.

(b) If Δt_m denotes the total time of flight of the projectile, the range Δx_m is given by $(v_0 \cos \theta)\Delta t_m$.

(c) The time Δt_p between the firing of the projectile and its arrival at point P is given by $(v_\mathrm{o} \sin \theta) / g$.

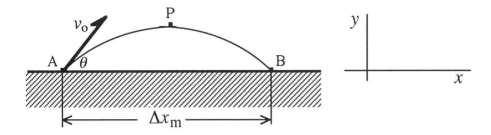

(d) The acceleration of the projectile has the constant magnitude g throughout the entire history of the flight.

(e) The magnitude of the velocity of the projectile is larger at point B than it is at point P.

(f) The magnitude of the velocity of the projectile is the same at point B as it is at point A.

(g) The momentum of the projectile is not conserved because the projectile does not constitute a closed system.

(h) The momentum vector of the projectile is rotating in the clockwise direction throughout the flight.

(i) The momentum vector of the projectile has its smallest magnitude at point P.

(j) The kinetic energy of the projectile changes during the course of the flight but is the same at point B as it was at point A.

(k) When the projectile strikes the ground at point B, its kinetic energy is entirely converted into thermal energy within the projectile and within the ground in the vicinity of the impact.

(l) None of the above.

CHAPTER 16

Naked Eye Astronomy

16.1 Describe how you would go about establishing vertical direction and horizontal direction, using the simplest possible materials or devices, at any place you happen to be located, especially in the case of a sloping hillside. (You are being asked, in other words, to give simple operational definitions of the terms "vertical" and "horizontal.")

16.2 Define the terms "local zenith point" and "local celestial meridian" and go on to do the following.

(a) Describe how you might go about identifying the direction of the local zenith point and mapping out the local celestial meridian at any place you happen to be located. Note that there are two different ways of doing the latter: one is to make use of the North Star, and the other is to make use of the shadow, cast by the sun, of a vertical stick. (It is legitimate to think of carrying out these tasks fairly crudely; high precision is not required.)

(b) Define the term "local noon" by describing how you would determine the moment of local noon without reference to a clock.

(c) Account for the fact that the actual moment of local noon varies continuously from east to west over any conventional time zone.

16.3 Give an operational definition of the term "geographic north-south direction" by describing how you would go about establishing this direction at any point at which you happen to be located.

(a) In the light of the universally accepted definition of "geographic north-south direction," explain in your own words why the magnetic compass does *not* serve to *define* this direction. What *is* the actual utility of the magnetic compass?

16.4 Note that there are two seemingly independent ways of determining the north-south direction at any location on the earth: (1) the direction, at the given location, to the pole star, and (2) the direction, at the given location, of the shortest shadow of a vertical stick, cast as the sun crosses the local celestial meridian.

(a) What significance do you see in the fact that these two entirely different observations give the same direction on the surface of the earth? That is, what does agreement between these two modes say about axis of rotation regardless of whether one thinks of the earth as rotating relative to the celestial sphere or the entire celestial sphere (including the sun) as rotating around the earth.

(b) Describe how a *different* arrangement of rotations (one that is *not* actually the case) might have led to disagreement between the two modes mentioned in part (a). (Hint: Consider the possibility of a diurnal motion of the sun different from what is actually observed.)

16.5 Naked eye observations of the sky lead very directly to the concepts of ''terrestrial and celestial poles'' and ''terrestrial and celestial equators.'' Describe in your own words how these concepts are formed; in other words, identify the *observations* that play the key roles and the *inferences* that are drawn from the observations.

16.6 We are looking vertically down on a sheet of paper with a stick mounted upright at its center (as shown by the small circle). Sketch on the diagram the shadows you would expect to see at your geographic location at the times of day requested below, on the date on which you are doing this problem. In each case choose a reasonable relative length and direction of the shadow, and label it with the letter designating the particular question.

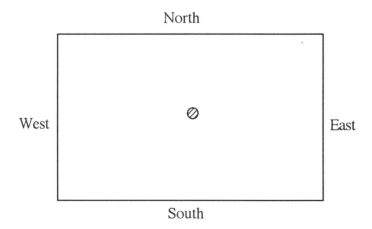

(a) Shortly after sunrise.   (d) Midafternoon.

(b) Midmorning.   (e) Shortly before sunset.

(c) Local noon.

16.7 Draw, for your geographical location, diagrams such as that required in Question 16.6 for the day of the winter solstice, the day of the vernal equinox, the day of the summer solstice, and the day of the autumnal equinox. Be sure to make the four diagrams consistent with respect to both direction and length of shadows at various times of day.

16.8 Suppose that on a particular day, you make observations of the shadow of a vertical stick from some time in late morning into early afternoon.

(a) A friend makes exactly similar observations on the same date at a location 2000 miles due east of yours. Do you expect the two records to differ in some substantial way or to be very nearly the same? If you expect them to differ, describe the difference you anticipate. In either case, explain your reasoning.

(b) Another friend makes similar observations at a point either due south or due north of your location (i.e., on the same meridian), but at the opposite latitude (i.e., on the other side of the equator). How do you expect this person's observations to differ from yours? Accompany your answer with appropriate sketches of shadows.

16.9 Consider the occurrence of a thin crescent moon:

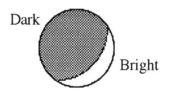

(a) How do you account for the dark (unilluminated) portion of the lunar disk?

(b) Can you remember under what circumstances of time of day or night (before or after sunrise, before or after sunset, before or after midnight, before or after noon) you are most likely to notice a thin crescent moon? If you can identify such a time, what relevance does it have to the answer you gave in part (a)?

16.10 (This is an exercise in arithmetical reasoning with the meaning of π and its relation to circular arcs.) The lunar orbit is approximately circular with a radius of 240,000 miles, and the moon shifts eastward through an angle of about 13° in one day.

  (a) Make a sketch of the situation just described, and, explaining each step of your reasoning carefully (do not simply substitute in formulas), calculate the distance the moon travels along its circular arc in one day. [Carry out your calculation in denary (powers of ten) notation.]

  (b) Compare this distance with some distances you know on the surface of the earth and interpret your comparison.

16.11 Suppose you are located in the northern hemisphere, looking south, and you see a moon half-illuminated, as shown, just crossing the local celestial meridian. When we speak of approximate times in the following questions, we refer not to clock hours but to times such as ''sunrise,'' ''shortly before or after sunset,'' ''midnight,'' etc. Be sure to explain your reasoning in answering each question.

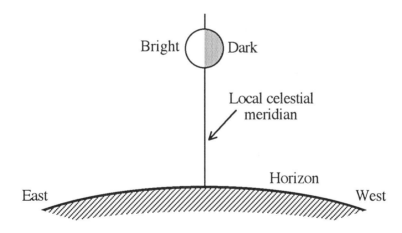

  (a) At approximately what time of day or night would you expect to see the configuration shown in the diagram? At approximately what time would this moon have risen? At approximately what time will it be setting?

  (b) Suppose you wish to look for the moon approximately 48 hours after the view represented in the diagram. Indicate roughly where in the diagram you would expect to see it, and sketch the approximate shape of the illumination you would expect to see.

16.12 Suppose that on a particular occasion, the full moon and a reference star are seen crossing the local celestial meridian simultaneously, as sketched in the diagram. Explain your reasoning in answering each of the following questions.

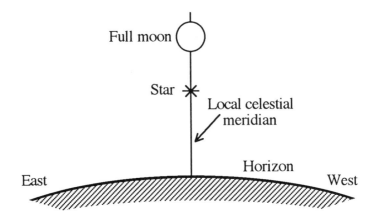

(a) What must be the approximate time of day or night?

(b) Sketch what the configuration above would look like approximately 3 hours later.

(c) Sketch the appearance (approximate phase) of the moon and the location of the moon when the same star crosses the local celestial meridian 3 or 4 days later.

16.13 In Samuel Taylor Coleridge's famous poem "The Rime of the Ancient Mariner," there occurs the following passage:

"Till clomb above the eastern bar
The horned Moon, with one bright star
Within the nether tip."

(a) These lines suggest the accompanying picture of the moon and neighboring stars. Do you see anything wrong with the "star within the nether tip"? Explain your reasoning.

(b) Is the picture of the moon correct for a northern or for a southern hemisphere viewer. Explain your reasoning.

16.14 What would be your response to a drawing in which a thin crescent moon is shown high in the sky around midnight? Explain your reasoning.

16.15 For simplicity, let us represent the earth as a perfect sphere. The poles and the equator are marked.

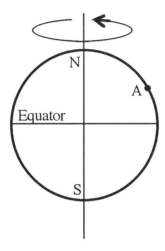

(a) Draw a line that shows the direction of the local zenith point for an observer located at point A, and label this line a.

(b) Draw a line representing the horizontal direction for an observer at point A, and label this line b.

(c) Draw a line that shows the direction along which the observer at A looks to see the North Star, and label this line c. (Remember that the North Star is extremely far away, a distance many millions of times the radius of the earth.)

(d) Assuming that the time is that of the vernal equinox and that the sun happens to be located in the plane of the drawing and off to the right, draw a line showing a ray of light from the sun arriving at point A, and label this line d. (Remember that the sun is more than 20,000 earth radii away from the earth.)

16.16 Show in careful detail how you would lead a fellow student to an understanding of the fact that the angular elevation of the North Star above our local horizon is numerically equal to the terrestrial latitude of our point of observation. Be sure to lead the student into defining the term "terrestrial latitude" and carefully drawing and labeling a relevant diagram.

16.17 Suppose you have available a piece of string, a weight, a soda straw, some pins, access to a telephone pole or other form of support, and a protractor. Describe how you might utilize these items to make a rough determination of your present latitude.

16.18 What simple observational evidence can you cite for the fact that the sun is not only farther away from us than the moon but is *very much* farther away?

16.19 Suppose a full moon happens to occur very close to the autumnal equinox. Where along your local horizon would you expect to see this full moon rising? Explain your reasoning with the help of a relevant diagram.

16.20 Let us consider some questions about the location of the sun relative to the local zenith point at noon at various places on the earth. In answering these questions, make use of appropriate diagrams and explain your reasoning.

(a) Define the term "ecliptic" and prepare to use it when relevant in the following questions.

(b) Does the sun *ever* pass through the local zenith point where you happen to live? Is it ever correct to say that "the sun is *overhead* at noon"? Why or why not?

(c) At what location on the earth does the sun pass directly overhead at noon at the time of the vernal equinox? Explain your reasoning with the help of relevant diagram.

(d) It happens that in San Juan, Puerto Rico, the sun actually does pass through the local zenith point at noon on two occasions: one some days *before* the summer solstice and the other some days *after* the summer solstice. Explain this observation with the help of a relevant diagram.

(e) What is meant by the terms "Tropic of Cancer" and "Tropic of Capricorn," and what band of behavior of the sun do they bracket on the surface of the earth?

(f) Suppose you were located close to the North Pole. Approximately where would you expect to see the sun at noon on the vernal equinox? Where would you expect to see it at noon a few days later?

(g) Answer question (f) from the point of view of an observer located simultaneously near the South Pole.

16.21 The diagram shows the earth at four different times of the year. An observer O is located at the equator, and the sun is off to the right in the plane of the diagram.

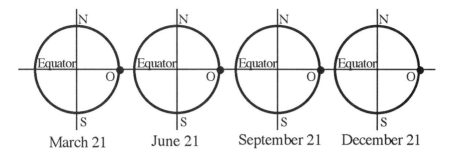

March 21   June 21   September 21   December 21

(a) In each of the four diagrams, draw four rays of light from the sun, incident on the earth, from the point of view of observer O.

(b) Describe where the sun is located at noon relative to O's zenith point on each of the four dates indicated. That is, is the sun *at* the zenith? North of the zenith? South of the zenith?

(c) Describe where the sun rises and sets, from O's point of view, on each of the four dates indicated. That is, does the sun rise north of east? Due east? South of east? etc.

(d) Describe the relative lengths of daylight and darkness from O's point of view on each of the dates indicated.

Note to the instructor: Question 16.22 can easily be reversed by giving the diagrams and asking for the corresponding dates.

16.22 In the world as we know it, it is an observed fact that the ecliptic and the celestial equator do *not* coincide. For the sake of this question, however, let us imagine a different world in which the ecliptic and celestial equator *do* coincide. In the following questions you are asked to predict how you would expect to see the sun behaving under these circumstances from various points of observation on the earth.

(a) On the diagram of the earth, show the location of the sun relative to the earth by drawing a set of rays coming to the earth from the sun, which is off to the right. (Remember that the sun is extremely far away relative to the size of the earth.)

(b) Consider the points of view of observers located at positions A, B, and C: Where would each observer see the sun rising and setting along the local horizon? At what elevation above the horizon would each observer see the sun crossing the local celestial meridian? How would these positions change during the course of the year? Explain your reasoning.

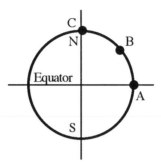

(c) Describe the relative duration of daylight and darkness for each of the three observers. How would these duration change during the course of the year? Explain your reasoning.

16.23 Roughly what path through the sky (location of points of rising and setting and elevation of point of crossing the local celestial meridian) would you expect the moon to follow at your location if a full moon happened to occur on the winter solstice? If it happened to occur on the summer solstice? Explain your reasoning.

16.24 Suppose the moon happens to be in its third-quarter phase at the time of the vernal equinox. Approximately where along the ecliptic must the moon be located? Explain your reasoning.

16.25 In this daytime view toward the south at some latitude in the northern hemisphere, the position of the sun is indicated.

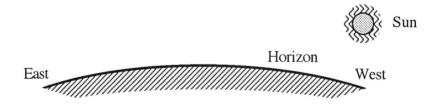

At the time being considered, Venus is known to be an "evening star." Mark on the diagram the position, relative to the sun, at which Venus must be located, and explain your reasoning.

16.26 The very bright star Sirius (sometimes called the Dog Star) follows Orion (the hunter) through the sky. For an observer at a latitude of 48°N, Sirius is observed to cross the local celestial meridian at an elevation of about 22° above the horizon. The diagram shows a cross section of the earth through the local celestial meridian for an observer at point A at latitude 48°N.

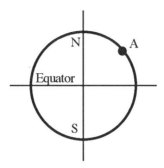

(a) At point A, draw and label the local horizontal and vertical lines.

(b) Using a protractor, draw and label a light ray arriving from Sirius at point A using the information given above. Mark angles appropriately.

(c) By adding appropriate lines to the diagram, determine whether Sirius is visible from the North Pole. If it is visible, is its highest elevation above the horizon greater or less than 22°? Explain your reasoning.

(d) By adding appropriate lines to the diagram, determine whether or not Sirius is visible from a point on the equator. If it is visible, is its highest elevation above the horizon greater or less than 22°? Explain your reasoning.

(e) During June, July, and, August, Sirius is not visible at all from point A at *night*. How do you interpret this observed fact?

16.27 In our study of elementary terrestrial physics, we gave the name "force" to any interaction that imparts acceleration to a material object, and we gave the name "gravity" to the interaction that accelerates material objects toward the earth.

(a) Describe in your own words how Newton enlarged and extended these concepts in creating a model for the solar system as a whole. Be sure to include the following ideas as well as others of your own: How do the concepts of "force" and "acceleration" get into the problem at all? What is meant by "extending" the idea of gravity? How does gravity get into the picture at all?

16.28 If we were to look down upon the plane of the solar system from a point above the northern hemisphere of the earth, we would see the earth revolving around the sun, the moon revolving around the earth, and the earth rotating on its axis—all in the counterclockwise direction as illustrated schematically. (The ellipticity of the earth's orbit and the length scales are greatly exaggerated.)

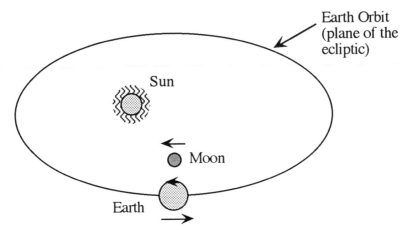

Earth Orbit (plane of the ecliptic)

Sun

Moon

Earth

Given the following definitions: (1) A *solar day* is the time interval for one rotation of the earth relative to the *sun*, i.e., the time interval between the sun's attaining its highest elevation above the horizon (local noon) at our point of observation. (2) A *sidereal day* is the time interval for one rotation of the earth relative to one of the fixed stars, i.e., the time interval for a star to return to the same position in the sky as it was observed to have on the preceding night. (3) A *synodic lunar month* is the time interval for the moon to return, in its revolution around the earth, to an initial position (or phase) relative to the sun, e.g., the time interval between exact full moons. (4) A *sidereal lunar month* is the time interval for the moon to return, in its revolution around the earth, to the position it initially had relative to some fixed star.

(a) Note that earth shifts its position in its revolution around the sun over the period of a month as well as over the period of a day. Note also the direction of this shift as well as the direction of the earth's rotation. In the light of these motions, how would you expect the length of a solar day to compare with the length of a sidereal day? That is, would it be longer, shorter, or of the same length? Explain your reasoning with the help of a diagram.

(b) It is an observed fact that the earth has a higher velocity in its elliptical orbit around the sun when it is near perigee (closest approach to the sun, occurring in December) than when it is near apogee (greatest distance from the sun, occurring in June). How would you expect the *difference* between the lengths of solar and sidereal days to change over the course of a year? Explain your reasoning with the help of a diagram.

(c) With respect to the relation between the synodic and sidereal lunar months, answer the questions that are asked in parts (a) and (b) with respect to the solar and sidereal days.

(d) In the light of the definitions given above, how might you expect the synodic and sidereal periods of the other *planets* to be defined? Explain with the help of a diagram.

16.29 How do you explain the fact that winters are much colder than summers in the northern hemisphere even though the earth is closer to the sun during the winter period than it is during the summer? Use diagrams to assist your explanation.

Note to the student: In the following multiple choice questions, circle the letters designating those statements that are <u>correct</u>. <u>Any</u> <u>number</u> of statements may be correct, <u>not</u> just one; you must examine each statement on its merits.

16.29 Understanding of a given set of ideas in science is frequently enhanced by carefully thinking through alternative situations, even ones that cannot possibly occur. The following question involves such thinking. We shall deal with the situation sketched in the figure: the earth is visualized as revolving around the sun in such a way that its axis of rotation is always tilted *toward* the sun. This is physically impossible because maintaining such a condition would require the presence of forces, acting on the earth, that do not exist in the solar system. In the actual situation, the axis of rotation of the earth maintains a very nearly fixed orientation relative to the fixed stars; it does undergo a relatively small motion, called "precession," which is exceedingly slow compared to the annual revolution around the sun, but it plays no significant role in our thinking about the effects that concern us in naked eye astronomy.

Let us contrast the situation illustrated here with what we know to be the case with respect to the revolution of the earth: the tilt of its axis of rotation relative to the plane of the ecliptic, the origin of the seasons, etc.

(a) Under these circumstances, it would be perpetual daylight at the North Pole and perpetual night at the South Pole.

(b) At the latitude at which you are located, days would always be longer than nights throughout the entire year.

(c) It would always be winter in the southern hemisphere.

(d) The positions of sunrise and sunset along the horizon (at any given latitude where the sun is observed to rise and set) would not change over the course of the year.

(e) At any one of the latitudes considered in part (d), the elevation of the sun would always be the same at local noon and would not change as the days go by.

(f) Instead of there being a fixed polestar, the location of the celestial pole would keep changing during the course of the year.

(g) The ecliptic (the apparent path of the sun against the field of stars) would not intersect the celestial equator.

(h) None of the above.

CHAPTER 17

Learning   Objectives

These statements of procedures and learning objectives were given to participants in summer institutes for high school physics teachers at the University of Washington at the start of each subject matter segment of the institute. Most of the participants worked in pairs in an essentially self-paced mode, proceeding through the text and laboratory work of either "PSSC Physics" or "Project Physics" according to their own choice. At the end of a segment, each participant had a conference with a member of the instructional staff, using the statement of learning objectives as a framework.

The conference was originally intended to be an oral examination that determined whether the participant had mastered the material of the unit at a sufficient level to warrant going on to the next subject matter unit. In practice, the conference turned out to be more an opportunity for the participant to synthesize and order his or her own knowledge rather than a test or examination. Some readers of these materials might find that they can be utilized (with suitable modifications and alterations) with students in introductory physics courses at either high school or college level. Some items may be useful as subjects for discussion in collaborative learning groups.

Unit 1. Kinematics

Participants should start right in on their respective curricula, studying text, working problems, and performing experiments. Working in pairs is encouraged. In general, it is advantageous to have someone to talk with about text and problems and to join forces in the performance of experiments. Learn to weave back and forth judiciously among these various activities according to the intent of the curricula themselves. For example, it is, in many instances, intended that laboratory experiments or observations *precede* reading of the text. Teachers who adhere to the structure and spirit of curricular materials in such respects will be better able to lead students in the manner intended by the progenitors of the curricula.

After you have completed this unit on kinematics, you should be able to do the following things:

(a) Give clear operational definitions of position s, change of position Δs, instant of time (or clock reading) t, interval of time Δt, *average velocity* $\bar{v}$, instantaneous velocity v, average acceleration $\bar{a}$, and instantaneous acceleration a.

(b) Translate the verbal description of a rectilinear motion into a graph of position versus clock reading or a graph of velocity versus clock reading (or both).

(c) Translate given position versus clock reading and velocity versus clock reading graphs into verbal descriptions of the motion represented or into a simulated motion with your own hand or body.

Note: The translations referred to in (b) and (c) can be enhanced and facilitated by use of the sonic range finder coupled to a microcomputer, displaying position, velocity and acceleration graphs of the motion of one's body.

(d) Work out representative problems at the end of each chapter, explaining your reasoning verbally in each case.

(e) Describe experiments by means of which you would be able to determine whether observed motions are uniformly accelerated, and be able to determine accelerations experimentally.

The following questions illustrate, but do not limit, some of the questions that might be asked of you in the exit interviews:

(a) What difficulties do your students have in acquiring an understanding of the concept of "instantaneous velocity," and how would you now go about helping them to overcome these difficulties?

(b) What is the meaning of the algebraic signs accompanying displacement, velocity, and acceleration in description of rectilinear motion? Explain in your own words how the algebraic signs get into the sequence in the first place. When we adopt a coordinate system and proceed to describe the full history of the rising and falling of a ball thrown vertically upward, does the acceleration of the ball change algebraic sign on the way down relative to the sign on the way up? Why or why not? Explain your reasoning in full detail, indicating how you would lead your own students into understanding what is involved.

(c) What is the meaning of $v = 0$ at the top of the vertical flight of the ball (or at the extreme end of the swing of a pendulum)? Is the object accelerating when $v = 0$? Is it wise to speak of the object as having "stopped" or "come to rest" at the given instant? Why or why not? What meaning do the terms "stopped" and "come to rest" convey to most people who have not studied physics? How do you propose to handle these matters with your students?

(d) Sketch a reasonable graph of the position versus clock reading history of several vertical bounces of a ball that has been dropped to the floor from rest at some initial height.

(e) A car accelerates from rest to a velocity $v = 80$ ft/s in 10 s. During this interval, it has traveled 500 ft. What can you infer about the acceleration in this case? Was the acceleration uniform or nonuniform? Explain your reasoning. Sketch what the v versus t history might have been like. If the distance traveled had been 400 ft instead of 500 ft, what conclusions could you reach about the nature of the acceleration?

Unit 2. Dynamics

After you have completed the unit on dynamics, you should be able to do the following things:

(a) Give clear, noncircular, operational definitions of the concepts of "force" and "mass."

(b) Define "centripetal acceleration" in circular motion and derive the expression for this acceleration in terms of radius and tangential velocity. (In this connection it also necessary to be able to define "tangential velocity.")

(c) Draw correct, well-separated force diagrams of *all* the objects that interact with each other in a given dynamical situation; describe each force in words (indicating the nature of the force and what object exerts it on what); and identify all Third Law pairs.

(d) Describe projectile motion as the superposition of two independent motions, one uniform and the other accelerated. Draw force, velocity, and acceleration diagrams for the projectile at any position in its trajectory.

(e) Work representative end-of-chapter problems in the texts.

The following questions illustrate, but do not limit, some of the questions that might be asked of you in the exit interviews.

(a) How would you discuss with a student the forces acting on an object resting on a table? (Be able to draw force diagrams for the object, the table, and the earth, indicating the relevant interactions.) What forces are acting on the object? What forces are acting on the table? What happens to these forces as you press down on the object with your hand? What happens to the table when you place the object on it in the first place? (That is, is the table deformed as a result of the force acting on it?) What is meant by the term "passive force" in this context? What happens as you increase the load on the table indefinitely? Is the table deformed or undeformed when you place a sheet of paper on it?

(b) Suppose a box rests on the floor and you push on it horizontally. Because of the presence of friction, the box does not slide immediately. Describe what happens to the frictional force as you start your horizontal force at zero and increase it until the point of sliding is reached. Draw a force diagram showing all the forces acting on the box and another diagram for the floor, showing the forces the box exerts on the floor. In what ways is this situation similar to that in the preceding question concerning the object placed on a table? In what ways is it different? Can friction be characterized as a "passive force"? Why or why not?

(c) Draw force diagrams for a cart on an inclined plank when the cart is held at rest and when it is freely accelerating down the plank. In each case also draw a force diagram for the plank. Explain the origin of the word "normal" in connection with the force exerted by the plank on the cart. How would you convince students that this force is indeed normal to the plank and is not directed vertically?

(d) A ball is placed in a cart or coaster wagon, and the cart is accelerated horizontally in a straight line. Describe carefully what you would see happening to the ball relative to the cart and relative to the ground, as you watched the experiment from the side. What is the point of leading students through such observations? In what sense is it *incorrect* to say that "the ball is thrown backward"? In what sense might this statement be correct?

(e) Discuss the floating and sinking of objects in water in terms of forces acting on the chunk of *water* the object will displace when it is immersed and then the forces that must act on the *object* once it has displaced the water. As an outgrowth of this discussion, give a simple nonmathematical justification of the assertion made in Archimedes's principle, namely, that an object in water is buoyed up by a force equal to the weight of water displaced.

(f) Suppose that a simple pendulum bob is suspended on a string from the roof of a car and from a rail at the periphery of a merry-go-round. How will the pendulum hang relative to its normal vertical orientation when the car moves at uniform velocity in a straight line? When the car is speeding up? Slowing down? When the car moves around a curve at constant speed? How will the pendulum on the merry-go-round hang when the latter is rotating? Draw separate force diagrams for the bob and the string in each case discussed.

(g) Suppose you are in an elevator that is in free fall relative to the earth. Describe what you would observe, relative to your frame of reference in the elevator, as you performed various experiments such as (1) letting a ball go from rest, (2) letting the ball go off with some low initial velocity in various directions such as horizontally or up or down, (3) bouncing the ball off wall, ceiling, or floor, (4) trying to set a pendulum swinging, (5) rotating a bob on a string, and (6) turning over a bucket of water. Make up some other possible experiments of your own.

(h) Consider the situation shown in the following diagram (you will surely recognize it as a widely used homework problem in the textbooks): the pulley and string have negligible mass; cart A rolls with negligible frictional resistance. The system is released from rest and accelerates as block B falls. We pose the following *qualitative* question: As the system accelerates, how does the force exerted on the cart by the string compare with the weight of block B? Is this force equal to, greater than or less than the weight of the block? Explain your reasoning *qualitatively*, showing how you make use of Newton's second and third laws to arrive at your conclusion without ''solving'' the problem analytically through the derivation of algebraic relations.

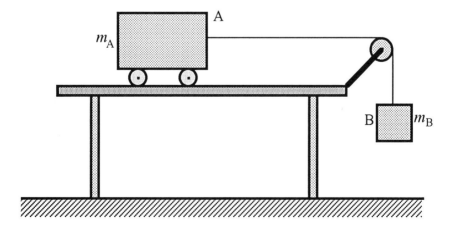

(i) Consider a pendulum bob suspended freely from the ceiling or some other support at each of three locations at the surface of the earth: the North Pole, the equator, and some intermediate latitude. Let us take the earth to be perfectly spherical (we know this is not actually the case), even though it is rotating. Sketch how the string on which the bob hangs would be oriented relative to a radial line from the center of the earth at each of the three locations if the earth were not rotating and with the earth rotating.

(j) Suppose we have a bob of mass m on a string and we set it rotating in a circle of radius R that lies in a *vertical* rather than a horizontal plane. Suppose we also manage to keep the tangential velocity v_{tan} constant throughout the circle (it is not important how this is achieved).

    (1) Consider the bob in two special instantaneous positions: one at the very top of the circle and one at the very bottom. Draw a force diagram for the bob in each case. Define carefully what is meant by the term "centripetal force" and argue that in these two cases, the centripetal force is *not* equal to the force T exerted by the string on the bob. What other force combines with T to make up the centripetal force?

    (2) Set up algebraic expressions for T at the top and bottom of the circle and interpret them in words, describing what happens to T as v_{tan} is varied from high to low values. How do you interpret the change of sign that T undergoes at the top of the circle in the case of such variation?

(k) Start with the situation in which a car goes around a curve on an unbanked road (i.e., the road surface lies in a horizontal plane). Draw a force diagram for the car, viewed from the rear, as it is going around the curve, being sure to show the location of the center of the curve in your diagram. What is the origin of the centripetal force that imparts the necessary centripetal acceleration?

    (1) Now assume that the road is *slightly* banked toward the center of the curve. Draw the force diagram of the car again, and explain why less demand is now placed on the radially directed frictional force on the tires than in the preceding case.

    (2) Explain what happens to various forces as the angle of banking is increased. How do you identify the angle at which there is no longer any demand placed on the frictional force? What happens if the angle of banking exceeds this value?

(3) Suppose we take a fixed angle of banking. Describe what happens on your force diagram as the speed of the car is increased from a very low to a very high value, straddling the optimum speed for which the road is banked.

(l) Be able to explain and interpret the following experiment that Newton describes in the *Principia*:

"I tried the thing in gold, silver, lead glass, sand, common salt, wood, water, and wheat. I provided two equal wooden boxes. I filled one with wood, and I suspended an equal weight of gold (exactly as I could) in the center of oscillation of the other. The boxes, hung by equal threads of 11 feet, made a couple of pendulums perfectly equal in weight and figure, and equally exposed to the resistance of the air. Placing the one by the other, I observed them to [swing] together forwards and backwards for a long while with equal vibrations. And therefore the quantity of matter [inertial mass] in the gold was to the quantity of matter in the wood as the action of the motive force [gravitational mass] upon all the gold to the action of the same upon all the wood; that is the weight of one to the weight of the other."

What was the point of this experiment? What did Newton observe? What inference is drawn from the results?

(m) Explain the basis for our belief that in the solar system, the earth and planets revolve around the sun rather than all the other members revolving around the earth.

Unit 3. Momentum and Energy

After completion of this unit, you should be able to do the following things.

(a) Give clear and detailed operational definitions of all the new terms that have been introduced such as "impulse," "momentum," "work," "kinetic energy," and "potential energy."

(b) Describe what happens in various commonly observed phenomena in terms of these new concepts. For example: describe, in terms of impulse delivered, momentum changes, work done, and various energy changes, what happens when a ball is thrown vertically upward and eventually ends up on the ground. Describe what happens when carts with spring bumpers collide and bounce apart. Describe what happens when a box is pushed along the floor so that it accelerates in the presence of friction, etc.

(c) Define the concept of "center of mass" and describe its role in phenomena involving momentum and energy transformations.

(d) Be able to solve both qualitative and quantitative end-of-chapter problems.

The following questions illustrate, but do not limit, some of the questions that might be asked of you in the exit interviews. (Be forewarned that these are sophisticated questions that are poorly handled in virtually all existing introductory texts.)

(a) Describe how you would lead your students, through appeal to common and easily observable phenomena, to distinguish between "temperature" and "heat," realizing, to begin with, that many students use these terms synonymously.

(b) Suppose you push a box along the floor at uniform velocity with a horizontal force P in the presence of friction. The frictional force f must be equal in magnitude and opposite in direction to P. In a displacement Δx you have done an amount of work $P\Delta x$. What has happened to this amount of work? Where, that is, has the energy gone? Is it correct to say that a negative amount of work $(-f\Delta x)$ has been done by the frictional force and therefore zero net work has been done on the block? Why or why not? (Remember that in physics we wish to define "work" as a form of energy put into, or taken out of, a system that has been specified—not just any occurring product of force and displacement.)

(c) In deriving the work–energy theorem, starting from $F_{net} = ma$, the displacement Δx that appears in the derivation must be that of the center of mass of the *particle* being accelerated. Thus the quantity of work done under these circumstances is the net force integrated over the displacement of the center of mass of the particle. Suppose we now proceed to compress a spring. Is the work done in compressing the spring the force exerted on the spring integrated over the displacement of the center of mass of the spring? What *is* the correct calculation? If we compress a gas in a cylinder by displacing a piston, is the work done on the gas equal to the force exerted by the piston integrated over the displacement of the center of mass of the gas?

(d) Suppose you stand on roller skates and push yourself away from a wall, leaving contact with the wall with kinetic energy of your entire body. Where has this energy come from? Did it come from work done by the wall on your body? Why or why not? Answer the same questions with respect to the situation in which you jump vertically upward.What is the source of the discrepancy we have run into in questions (c) and (d)? What are some basic limitations we must place on applicability of the work–energy theorem derived from $F_{net} = ma$?

Unit 4. Electricity and Magnetism

After you have studied the text material on electricity and magnetism and have done representative homework problems, you should be able to give clear and specific answers to the following questions in exit interviews.

(a) In operational terms, under what circumstances do we describe an object as being "electrically charged"? What is meant by the term "electrical charge"? What can one say about what electrical charge "is"? What observations lead us to visualize electrical charge as *transferable* from one object to another?

(b) In electrostatic situations, what observations lead us to characterize some substances as "conductors" of electric charge and other substances as "nonconductors"?

(c) What observations lead to characterizing two objects as carrying "like" charges and to the assertion that "like charges repel? That is, does introduction of the term "like" stem from arbitrary definition or from experimental observations?

(d) What is meant by the term "unlike charges"? Use of only two terms (like and unlike, positive and negative) implies that there are no more than two varieties of electrical charge. What leads us to accept this idea? (What experimental observation would force us to recognize a third variety if it should turn up?)

(e) Note that electrical charge is *not* a material substance and cannot be measured by its mass or any other such "tangible" property. Through what sort of experimental observations is it possible to assign numerical values to "quantity of electrical charge"? What observable effects, that is, can be used to help quantify electrical charge?

(f) State Coulomb's law clearly and precisely, defining all terms occurring in the algebraic statement , and indicating to what situations this law is directly applicable as well as at least one or two situations in which it would *not* be directly applicable. (Is the law directly applicable, for example, to the case of two charged rods attracting or repelling each other?)

(g) It is an observed fact that a charged object carrying either variety of electrical charge attracts any uncharged object. What explanation is invented for this occurrence? What prior observations and experiences motivate the construction of this model (the model to which we give the name "polarization")?

(h) Consider the following situations in which there is a clear interaction between an electrically charged object and an uncharged object: (1) a charged rod attracts an uncharged pith ball or a bit of paper; (2) a charged rod charges, on contact, a pith ball covered with a conducting coating; (3) the leaves of an electroscope separate when a charged rod is brought near the electroscope without actually making contact; and (4) an electroscope is charged by induction.

> (1) In each case be sure that you are able to describe, in terms of any one of the three following models and with the aid of clear diagrams, the interactions that take place: (i) negative charge is taken to be stationary while positive charge is free to move; (ii) positive charge is taken to be stationary while negative charge is free to move; and (iii) both varieties of charge are free to move. Argue, in your own words, that at the level of macroscopic observations, it is impossible to confirm or to rule out any of these three models.

(i) Consider the following sets of interactions among various objects — interactions to which we have already given different names: (1) electrostatic interactions of charged bodies with each other and with uncharged bodies, (2) magnetic interactions of permanent magnets with each other and with unmagnetized pieces of iron such as nails or paper clips, and (3) gravitational interactions among all material bodies in the universe.

> (1) Describe as many observations as you can that show there to be profound differences among the three types of interaction and argue that these differences provide the justification for recognizing three different phenomena and therefore for creating three different names.

(j) If, in your studies, you introduced the term "electricity" initially in connection with what we call "electrostatic interactions," describe experiments and observations that justify the conclusion that the phenomena associated with what we call "electrical batteries" and household "electrical outlets" are related to electrostatic interactions in such a way that it is legitimate to invoke the term "electrical" for all these seemingly very different phenomena. If, on the other hand, you started with what we call "current electricity," describe observations and experiments that reverse the preceding line of argument and justify using the word "electrical" in connection with interactions among objects that have been rubbed with other materials.

(k) Be able to sketch electrostatic force patterns (i.e., "electrical lines of force") in the neighborhood of various simple configurations such as: a point charge, a line charge, relatively close to a charged plane, a uniformly charged sphere, charged capacitor plates, two like point charges, and two unlike point charges. (Be sure that you know the convention for establishing the *direction* of the field at any point, not just the overall pattern.)

(l) Be able to account for the fact that except for small fringe effects near the edges, the electrical field between capacitor plates is confined to the region between the plates with zero field outside, while the electrical field of a single charged plate extends to great distances (relative to the size of the plate) on either side.

(m) Be able to sketch the magnetic force patterns (i.e., "magnetic lines of force") around such simple magnetic and electromagnetic configurations as a single magnet, between like and unlike magnetic poles, around a current-carrying wire, around a coil of wire, and around a solenoid. (Be sure that you know the convention for establishing the *direction* of the field at any point, not just the overall pattern.)

(n) Be able to predict the direction of force on an element of current-carrying wire at any location in a magnetic field of given direction.

(o) Be able to indicate the direction of the force (i.e., the "Lorentz force") that would act on either a positive or negative electrically charged particle moving in any given direction at any specified location in a magnetic field.

(p) Two parallel wires carrying electric current in the same direction are observed to attract each other. Following Ampere, we call this interaction "electromagnetic." Since the wires are connected to a battery or to some other electrical source, the interaction might have been electrostatic. How do we know that it *not* electrostatic? That is, what observations and experiments compellingly indicate that the interaction differs completely from electrostatic interactions.

Unit 5. Wave Phenomena

I. After you have studied the behavior of waves in one dimension (e.g., waves on a stretched string and on a spring such as a "slinky") you should be able to do the following.

(a) Describe, in your own words, without use of any new technical terms, what is meant by the terms "wave pulse" and "wave train." In developing your description, make clear what behaviors distinguish wave motion from the ordinary particle motion studied earlier. Is particle motion still present in the wave phenomena you are now describing? If so, in what way? (Be sure to sketch relevant pictures in the course of your description.)

(b) Describe clearly and precisely in ordinary language (i.e., with a minimum of technical terms) and with the help of relevant diagrams the differences between transverse waves on the one hand and longitudinal waves on the other.

(c) Define the term "propagation velocity, V" of a pulse and describe what factor or factors determine or affect this velocity in transverse and longitudinal pulses in various media.

(d) Define the term "particle velocity" in a wave pulse and point out the relevant motions in actual transverse and longitudinal pulses that you initiate on strings and slinkies. How is the particle velocity directed (relative to the direction of wave propagation) in the case of a longitudinal compression pulse? In a longitudinal rarefaction pulse?

(e) Describe, sketch, and interpret graphs that might be used to represent propagating *transverse* wave pulses. (Note that there are two possible variables for the abscissa: clock reading t and position x along the medium. There are two possible ordinates: transverse particle displacement y from equilibrium position and transverse particle velocity v_y in the y direction. You should be able to use and interpret any of the possible representations.) Interpret the meaning of the algebraic signs that accompany the variable y (i.e., how do these signs originate and how are they to be interpreted physically?). Define the term "amplitude" in each representation.

(f) Given a stretched string, describe how you would generate (1) a transverse pulse with only positive deflection, i.e., only a positive phase; (2) a transverse pulse with only negative deflection, i.e., only a negative phase; and (3) a transverse pulse with both positive and negative deflections, i.e., with both positive and negative phases. (In your description, indicate what motion you would actually execute with your hand at one end of the string.)

(g) Describe and interpret graphs that might be used to represent propagating *longitudinal* wave pulses. What possibilities are there for the ordinate? (Note that in addition to particle displacements and velocities, there is the possibility of defining a "density" of the medium that could serve as a useful variable.) Interpret the meaning of the algebraic signs that accompany the variables you have chosen as ordinates for your graphs. How do these signs originate and how are they to be interpreted physically? What is meant by the terms "compression" and "rarefaction"? Define the term "amplitude" in each representation.

(h) Given a stretched coil spring (say, lying on the table), describe how you might generate (1) a longitudinal pulse with only a positive phase, (2) a longitudinal pulse with only a negative phase, and (3) a longitudinal pulse with both positive and negative phases. (In your description, indicate what motion you would actually execute with your hand at one end of the spring.) Now return to the description you gave in item (f) above, and identify the very significant differences between what happens in the case of transverse waves on the one hand and longitudinal waves on the other when it comes to generating pulses having only one phase, positive or negative.

(i) Describe what happens when transverse and longitudinal wave pulses traveling in opposite directions "collide." How does the behavior differ from what happens when particles collide?

(j) Define what is meant by "superposition" of wave shapes, and describe what happens when superposition occurs with either transverse or longitudinal pulses. (You should be able to describe superposition either in terms of particle displacements or particle velocities.)

(k) Define, using words and diagrams, "constructive interference" and "destructive interference" in terms of superposition of wave pulses.

(l) Given the shape of a transverse or longitudinal pulse incident at a fixed or at a free boundary, be able to predict the shape of the reflected pulse. (You should be able to deal with pulses of asymmetric shape, either positive or negative, either compressions or rarefactions. You should be able to make such predictions through visualizing the reflected pulse as emerging from the fictitious region beyond the boundary and having a shape such that, when superposed on the incident pulse, the layer of medium at the boundary maintains its fixed or free character. Maintaining the character of the boundary is called "satisfying the boundary conditions.")

(m) Describe what is meant by a "continuous wave train" in contrast to a single pulse. Describe what is meant by a "periodic" wave train. Describe what is meant by a "sinusoidal" wave train. Describe how you would generate transverse and longitudinal periodic wave trains in the media with which you have worked. Define and illustrate with appropriate diagrams what is meant by "constructive" and "destructive" interference of periodic wave trains.

(n) With the help of appropriate diagrams, define the terms "frequency, f" and "wavelength, λ" of a periodic wave train. Then, reasoning *arithmetically* from the definitions, establish the relationship among the three quantities V (velocity of propagation), f, and λ. (In other words, you should be able to reason out the relationship whenever you need it rather than memorizing it as a rigid formula.)

II. These learning objectives concern the behavior of straight and circular wave pulses (not wave trains) generated in a ripple tank. After studying the behavior of such pulses, you should be able to do the following.

(a) Define the concepts "ray" and "wave front," using diagrams as well as words, and illustrating the concepts in the cases of both straight and circular pulses.

(b) Using appropriate diagrams, define the concepts "angle of incidence" and "angle of reflection" for both straight and circular pulses incident at a straight rigid barrier. Define "normal" and "glancing" incidence. Be able to sketch these angles in wave front as well as ray representations. Be able to sketch what happens as the angle of incidence is increased from normal to glancing.

(c) Suppose a straight wave pulse propagating in a region of deeper water (higher propagation velocity) is incident at a straight interface with a region of shallower water (lower propagation velocity). Sketch separate ray and wave front diagrams showing the incident, transmitted and reflected pulses. Sketch such diagrams for the case in which the situation is reversed and the incident wave pulse arrives in the shallower region. In connection with your diagrams, define the concept "angle of refraction" (or "angle of transmission").

(d) In each of the instances and diagrams arising in item (c), be able to sketch how the angle of refraction changes as the angle of incidence is varied from normal to glancing.

(e) On the basis of your observations with circular wave pulses, be able to sketch what happens in reflections from a straight rigid barrier; i.e., show how the reflected wave changes as you move the center of the incident wave closer to, or farther from, the barrier.

(f) Sketch what happens when circular pulses are incident at a straight refracting interface, making diagrams that show different distances of the center of the circular pulses from the interface.

(g) In moving an object, say a pencil, through a ripple tank, describe the circumstances in which a bow wave is *not* formed by the moving pencil and the circumstances in which a bow wave *is* formed. Explain how the two situations differ.

III. These learning objectives concern the behavior of periodic wave trains in a ripple tank.

(a) Sketch the wave front patterns observed in the ripple tank when straight wave trains are incident at a straight rigid barrier at different angles of incidence.

(b) Sketch the wave front patterns observed when straight wave trains are incident at a refracting interface at different angles of incidence. (Be able to do this for incidence in both the deeper and the shallower water regions.)

(c) A straight wave train arrives at normal incidence to a straight barrier that is shorter than the length of the wave fronts (i.e., the unimpeded portion of the wave front can propagate past the barrier while part of the wave front is blocked). Sketch the pattern to be observed in the region beyond the barrier for the cases in which the wavelength λ is very short relative to the length of the barrier, very long relative to the length of the barrier, and of intermediate length.

(d) A straight wave train arrives at normal incidence to a straight barrier that contains an opening of width D, and the waves are blocked except for passage through the opening. Sketch the pattern of wave fronts transmitted through the opening for different ranges of the ratio of wavelength to opening width λ/D, i.e., for small, large, and intermediate values of λ/D.

(e) Explain why the *wavelength* of a wave train changes when the train is transmitted through a refracting boundary while the frequency of the train remains unchanged.

(f) Circular wave trains of wavelength λ, from two point sources running in synchronism a distance d apart, form an interference pattern. Sketch the pattern for different values of d at a fixed value of λ and for different values of λ at a fixed value of d. Identify the regions of constructive and destructive interference in each pattern. Selecting any arbitrary location in any one of the patterns, be able to say how many wavelengths the wave front from one of the sources leads or lags the corresponding (simultaneously emitted) wave front from the other source.

(g) In the light of the ideas that have been developed, define the concepts ''refraction,'' ''diffraction,'' and ''interference'' and distinguish among them.

IV. The following learning objectives involve the experience of transferring, in the abstract, what has been learned about waves on coil springs and in ripple tanks to the phenomenon of sound waves.

(a) In the light of what you have seen of the generation and behavior of compression and rarefaction pulses on a coil spring, sketch what you visualize might be happening to the air in a tube when a piston or diaphragm moves rapidly back and forth at one end.

(b) What variables (ordinate and abscissa) might you use to make a graph of a sound pulse?

(c) Is it possible to make a sound pulse having only a compression and no rarefaction phase by moving the piston inward and returning it to its initial position? Why or why not? How does this situation compare with the generation of pulses on the coil spring?

(d) How would you imagine the interference of sound waves to take place? Suppose you had two audible point sources of sound (analogous to the situation with two point sources in the ripple tank), e.g., two tuning forks sounding in unison. How would you go about finding regions of constructive and destructive interference?

(e) Use what you have said in items (d) and (l) in Section I above to predict how the compression and rarefaction phases of sound pulses would be reflected from a rigid wall.

Unit 6. The Nature of Light

After studying the material concerning the model for the behavior of light that is accepted in classical physics, you should be able to do the following.

(a) Describe experiments in which the behavior of light can be consistently accounted for on the basis of a corpuscular (i.e., particle) model.

(b) Describe experiments in which a corpuscular model *fails* to provide a consistent model for the behavior of light and explain in what *way* the corpuscular model fails.

(c) Describe in what way the wave model is successful where the corpuscular model fails and therefore why the wave model is ultimately accepted.

(d) Describe how the wave model accounts for the observed fact that the center of the Newton's rings experiment (where the thickness of the air film between the two pieces of glass is very much smaller than the wavelength of light) is dark rather than bright.

(e) Describe experiments that indicate light waves to be transverse rather than longitudinal, even when one still does not know the wave to be electromagnetic in nature.

(f) Describe commonly observed phenomena that hint (not necessarily prove) that light might be intimately connected with electric and magnetic effects on the microscopic level of the structure of matter.

CHAPTER 18

Term Paper Assignments

Term paper assignments can have a variety of desirable goals, and each teacher will emphasize different priorities. The following sample assignments have the principal goal of cultivating aspects of "scientific literacy" such as those defined in Chapter 12 of *A Guide to Introductory Physics Teaching.*

   Vaguely stated term paper assignments are not, in general, very effective for students in introductory courses. Without at least some guidance toward ways of focussing discussion and development, only a very few students are likely to write papers that penetrate as deeply as many are fully capable of doing. The problem for the instructor is to provide enough guidance to help focus attention on significant issues without imposing excessive constraint or specifying desired conclusions. Experience with such assignments indicates that guidance is most fruitful when it marks out degrees of specificity in the discussion being called for but allows substantial range of choice as to examples to be used and ideas to be analyzed.

   It is also fruitful to lead students into describing and assessing their personal learning experiences in the given context. At this stage very few students, if any, have had the opportunity to assess what it means to have learned and understood a significant range of scientific ideas, and even fewer have had the opportunity to describe a learning experience of their own in their own words. Given such opportunity for reflection and assessment, many students articulate penetrating insights that do not emerge in other circumstances.

   Following are a few examples of term paper assignments the author has used, with varying degrees of success and with various student populations. Most have to do with aspects of the Newtonian synthesis, but one has to do with the concept of the "electron." The intent here is not to promulgate specific assignments but rather to provide at least one model for the construction of term paper assignments that enhance student response and performance. A teacher should, in the final analysis, generate his or her own assignments geared to the level of readiness and the vocabulary of the students.

Example 1

Write a coherent essay, approximately 1000 words in length, describing the evolution of the law of universal gravitation as we studied it in its historical context and discussing the alteration it represented in contemporary points of view toward terrestrial and celestial phenomena. Include examples of calculations that you yourself can make as a result of what you have been learning. You might, if you wish, include a discussion of the contribution made by Cavendish (the experiment he described as "weighing the earth" and which we interpreted as measuring the universal constant G). If you elect to do so, be sure to indicate when, in the historical sequence, this work was done.

In the course of your discussion, show that you understand the meaning of the term "empirical information" in connection with Kepler's laws and the relevance of this idea to the algebraic development of the gravitation law. Also show awareness of the distinction between inductive and deductive reasoning and point out specifically, as you go along, what kind of reasoning is being described at any given point.

In illustrating calculations you yourself can make, you are free to choose problems from the textbook or to make up a problem or problems of your own (special credit will be given for the latter mode). You might calculate the tangential velocity of a satellite in near-earth orbit and compare it with values that have been quoted in newspapers or magazines. In any case, in working out any problem, explain all steps as an author would in a textbook, and interpret your results. Worked-out problems should, however, be incorporated smoothly in the story line of your paper, not presented as abrupt, isolated entities.

Example 2

The following quotation is from *Patterns of Discovery* by N. R. Hanson (Cambridge University Press, Cambridge, 1958).

"[The formula] $F = Gm_1m_2/R^2$ did remarkable work in the 'Principia'. . . for the law unified the laws of Kepler and Galileo into a powerful pattern of explanation—one of the most powerful in the history of physics. For Newton the law . . . did not simply 'cap' a cluster of prior observations: it did not summarize them. Rather it was discovered as that from which the observations would become explicable as a matter of course. Newton was not an actuary who could squeeze a functional relationship out of a column of data; he was an inspired detective who, from a set of apparently disconnected events (a bark, a foot print, a 'faux pas,' a stain) concludes, 'The game keeper did it.' No one less than a Newton, given the laws of Galileo and Kepler, observations of the lunar motion, the tides, and the behavior of falling bodies, could infer that

$F = Gm_1m_2/R^2$. This law organizes and patterns all those things, and others as well, but nothing incompatible with any of them.

"The conceptual situation is not unlike this: novel mathematical theorems are encountered which, besides being individually surprising, do not seem to fit together as a system . . . . Philosophers sometimes regard physics as a kind of mathematical photography and its laws as formal pictures of regularities. But the physicist often seeks not a general description of what [he or she] observes, but a general pattern of phenomena within which what [is observed] will appear intelligible. It is thus that observations come to cohere systematically . . . . "

Write a paper of approximately 1500 words using the quotation above as a point of departure for discussion of what you learned in studying the Newtonian synthesis. (Note that the Newtonian synthesis includes the laws of dynamics as well as the law of gravitation.) You might want to consider and deal with at least some of the following questions. It is up to you to decide what to write; it is not possible to cover all the questions suggested.

(a) Are you surprised by any of Hanson's remarks? Do you agree or disagree with him? (There is ample room for disagreement. You are not obligated to accept everything he says.) Why do you think Hanson says that the law of gravitation did "not simply 'cap' a cluster of observations"? In what sense is it reasonable to say that Newton was an "inspired detective" rather than an "actuary"? (If need be, look up the meaning of the latter word.)

(b) In what sense is the word "explanation" used in the context under consideration? What do you see to have been "explained" by the Newtonian synthesis? What do you see as *not* explained? Has $F = ma$ been "explained"? What relevance do you see in Newton's famous remark: "But hitherto I have not been able to discover the cause of those properties of gravity from phenomena [i.e., observation and experimentation], and I frame no hypothesis . . . . To us it is enough that gravity does really exist, and act according to the laws which we have explained, and abundantly serves to account for all the motions of the celestial bodies and of our sea." (It should be noted that to Newton and his contemporaries, the term "hypothesis" referred to mystical or occult ideas that were to be avoided in natural philosophy. Thus it carried a pejorative connotation—not a favorable and respectable one as it does at the present time.)

(c) How do *both* m_1 and m_2 get into the gravitation formula? What is the connection between "mass" in this context and inertial mass in $F = ma$?

(d) Look up the meaning of the words "inductive" and "deductive" as used in connection with scientific reasoning and list at least three or four specific instances of each type of reasoning in the sequence you studied. What do these instances have to do with Newton's "detective work"?

(e) Is a philosopher wrong in regarding physics as a "a kind of mathematical photography"? What do you think Hanson means in the sentence that contains this metaphor? What is meant by "observations come to cohere systematically"? Do you think that the basic laws or formulas "come to cohere systematically" in the same sense as do the observations to which the formulas apply? Why or why not?

(f) Has your study of this episode in the history of ideas altered or expanded any of your notions of the nature of science? If so, in what way? When did the alteration occur? What have you learned aside from "physical facts"? What is your present view of what it was that *Newton* learned?

Example 3

[The Austrian physicist Ernst Mach, who was a professor of physics at the University of Prague, was one of the great physicist–philosophers of the nineteenth century. In 1883, in his book titled *The Science of Mechanics*, he subjected the Newtonian theory to a most sophisticated and searching criticism. With judgments, insights, and perspectives inherited from the two hundred intervening years of scientific thought, he devastatingly analyzed Newton's sometimes circular definitions, fallacious justifications of the concepts of absolute space and time, and repeated appeal to "unscientific hypotheses" (which Newton professed to avoid but slipped into unwittingly). Mach provided an important part of the criticism and analysis that ultimately led to Einstein's reexamination of the foundations of the entire theoretical structure and his formulation of the theory of relativity. We shall not concern ourselves with Mach's critical analysis but shall use as a keynote some of his remarks about the Newtonian synthesis in general. The following passage is taken from the English translation of the second German edition of *The Science of Mechanics*. The Open Court Publishing Company, Chicago, 1893.]

"But in addition to the *intellectual* performance [in the *Principia*], the way to which was fully prepared by Kepler, Galileo, and Huygens, still another achievement of Newton remains to be estimated, which in no respect should be underrated. This is an achievement of *imagination*. . . . Of what nature is the acceleration that conditions the curvilinear motion of the planets . . . ? Newton perceived, with great audacity of thought, and first [according to his own reminiscences] in the instance of the moon, that this acceleration differed in no substantial respect from the acceleration of gravity so familiar to us. It was probably the principle of continuity, which accomplished so much in Galileo's case, that led him to his discovery. He was wont . . . to adhere as closely as possible, even in cases presenting altered conditions, to a conception once formed, to preserve the same uniformity in his conceptions that nature teaches us to see in her processes . . . . The motion of the moon thus suddenly appeared to him in an entirely new light, but withal under quite familiar points of view . . . . The new conception was attractive in that it embraced objects that previously were very remote, and it was convincing in that it involved the most familiar elements. This explains its prompt application in other fields and the sweeping character of its results . . . . Thus an amplitude and freedom of physical view were reached of which men had no conception previously to Newton's time."

Write a paper of approximately 1500 words using the quotation above as a point of departure for discussion of what you learned in studying the Newtonian synthesis. (Note that the Newtonian synthesis includes the laws of dynamics as well as the law of gravitation.) You might wish to amplify some of Mach's comments with specific illustrations. You might want to consider and deal with at least some of the following questions. It is up to you to decide what to write; it is not possible to cover all the questions suggested.

(a) In what sense is the term "synthesis" used in this context? What was it that Newton "synthesized"? What aspects did the synthesis *not* include? Be sure to give specific examples.

(b) In what sense did the motion of the moon appear "in an entirely new light"? Did you have a similar personal experience with aspects of the ideas being studied? If you did, try to identify and describe the nature of this experience.

(c) What was the "principle of continuity that accomplished so much in Galileo's case"? (Galileo, for example, argued that the restrained behavior of the ball rolling down the inclined trough in slower, more readily measurable motion, should be related to its behavior in free fall.) What arguments based on such uniformity in nature have you encountered up to this point? Give several specific illustrations of Newton's use of arguments based on continuity and uniformity among apparently disparate natural phenomena.

(d) If you have an interest in intellectual history, illustrate and expand on the consequences alluded to in the last sentence of the quotation.

A note of information: Newton devoted considerable time and effort to accurate measurements of the period of a pendulum consisting of a hollow housing into which he placed various materials differing in chemical nature and composition. From these measurements he calculated the value of acceleration due to gravity, g, showing it to be independent of the nature of the materials. Why do you think he did these experiments? You might find this episode relevant to dealing with some of the questions posed above.

In writing this paper, you are not being asked to become a historian of science and discover lines of reasoning by which Newton himself arrived at various insights. Such specific details are hidden for the most part even from professional historians. Newton published in what was to him the final, polished, convincing form and allowed very little of his "private science" to become known to others. Your task is simply to *interpret* and *illustrate*, thereby deepening your own understanding of this dramatic incident in our cultural heritage.

Example 4

The following quotation is taken from *The Mechanization of the World Picture* by the Dutch historian of science E. J. Dijksterhuis (Clarendon Press, Oxford, 1961).

". . . [P]eople had always speculated about the movements performed by material bodies under the influence of internal or external causes; for this purpose they had used various terms, which, because they also occurred in everyday speech, appeared quite clear, but which in reality were not defined sharply enough to be simply or safely employed in scientific discussions. Scientists had spoken of gravity, levity, force, power, velocity, resistance, tendencies, sympathy, antipathy, impetus, quantity of motion, mass, centrifugal force . . . , and the force of an impact, without ever having adequately defined any of these concepts. Without formulating them explicitly, scholars had started from certain general notions, which had been borrowed from day-to-day experience and therefore appeared to be evident, but afterwards all these notions were found too inadequate for an exact treatment of the subject to be based on them. Gradually, indeed, doubt had arisen as to the correctness of the . . . dynamics founded on these notions; in particular a new notion of inertia had replaced the old one. But while this was going on, other conceptions from ancient dynamics, specifically the proportionality of force and velocity, had been preserved in full, and scientists had always omitted to define accurately the numerous terms they used...

"It was Newton's task to create order in this chaos of terms and notions. The best method would have been that of Hercules cleansing the Augean stables, i.e., radical rejection of the old and subsequent reconstruction from the bottom. In this case it would have meant placing mechanics on a new foundation with the aid of sharply defined terms, preferably not taken from everyday speech, so that they were not yet charged with misleading associations. But science in its actual development is not accustomed to heed such semantic ideals: any man who tries to reorganize it has himself grown up in the world of thought he wishes to reform, thinks in its concepts, and speaks in its terms . . . ''

Use this quotation as a point of departure for discussion of your own learning experience in studying Newtonian mechanics. Fulsome adulation of Newton and uncritical acceptance of every remark of a commentator are unnecessary. There is ample room for agreement and disagreement, for modifications of point of view, and for acknowledgment of ignorance. Think through your ideas and your experience carefully, and, drawing on your present level of knowledge, present them with some courage of conviction, using specific illustrations to support your statements.

You might wish to consider and enlarge on some of the following questions.

(a) Has you own view of the meaning of terms such as "velocity," "acceleration," "force," "mass," "inertia," and "gravity" changed in the course of this study? If so, at what point and in what ways?

(b) What do you see to be the role of language and the influence of everyday speech in clouding or illuminating formation of scientific insights? Do you think it might have been possible for Newton to have "cleansed the Augean stables"? If so, how might he have done so?

(c) In the light of your experience so far, what problems can you begin to anticipate in connection with introduction of terms such as "energy," "work," "heat," and "power" in forthcoming discussions?

(d) In the light of your own study of this episode in the history of science and ideas, do you agree that Newton brought "order out of chaos?" Why or why not?

(e) If you have taken courses in English literature, have you encountered any analogous problems in connection with interpretation of a literary work?

(f) Have you encountered references to Newton or the the Newtonian synthesis in any of your courses in literature or history? If so, you might wish connect some of these references with what you have to say in this paper. Such connections deserve a bonus in grading.

Example 5

Note to the instructor: The first part of this assignment should be made *prior* to study of the Thomson experiment on charge-to-mass ratio of entities in the cathode beam as outlined in Chapter 10 of *A Guide to Introductory Physics Teaching*.

Write a brief (one-page) paper addressed to the following questions: What does the term "electron" mean to you at the present time? How do we come to know about such an entity? What evidence is there for its existence? What are some of its properties? (Submit one copy of the paper to your instructor and keep one copy for your own use in connection with the second part of this assignment.)

Note to the instructor: The second part of the assignment should be made *after* study of Thomson's experiments on the cathode beam as outlined in Chapter 10 of *A Guide to Introductory Physics Teaching*.

In the light of your study and interpretation of the Thomson experiment, address the questions about the concept of "electron" that were raised in the first part of this assignment. Compare your present view of the concept with that which you held prior to this study. In what ways, if any, have your insight and understanding changed? Describe in detail, using specific examples. What do you believe it means to "understand" such a scientific concept?